Mobilizing Science-Based Enterprises for Energy, Water, and Medicines in Nigeria

Committee on Creation of Science-Based Industries in Developing Countries

Development, Security, and Cooperation

Policy and Global Affairs

NATIONAL RESEARCH COUNCIL
OF THE NATIONAL ACADEMIES

NIGERIAN ACADEMY OF SCIENCE

THE NATIONAL ACADEMIES PRESS
Washington, D.C.
www.nap.edu

THE NATIONAL ACADEMIES PRESS 500 Fifth Street, N.W. Washington, DC 20001

NOTICE: The project that is the subject of this report was approved by the Governing Board of the National Research Council, whose members are drawn from the councils of the National Academy of Sciences, the National Academy of Engineering, and the Institute of Medicine. The members of the committee responsible for the report were chosen for their special competences and with regard for appropriate balance.

This study was supported by funding from the National Academies. Any opinions, findings, conclusions, or recommendations expressed in this publication are those of the author(s) and do not necessarily reflect the views of the organizations or agencies that provided support for the project.

International Standard Book Number-13: 978-0-309-11118-8
International Standard Book Number-10: 0-309-11118-8

Additional copies of this report are available from the National Academies Press, 500 Fifth Street, N.W., Lockbox 285, Washington, DC 20055; (800) 624-6242 or (202) 334-3313 (in the Washington metropolitan area); Internet, http://www.nap.edu.

Printed in the United States of America

THE NATIONAL ACADEMIES

Advisers to the Nation on Science, Engineering, and Medicine

The **National Academy of Sciences** is a private, nonprofit, self-perpetuating society of distinguished scholars engaged in scientific and engineering research, dedicated to the furtherance of science and technology and to their use for the general welfare. Upon the authority of the charter granted to it by the Congress in 1863, the Academy has a mandate that requires it to advise the federal government on scientific and technical matters. Dr. Ralph J. Cicerone is president of the National Academy of Sciences.

The **National Academy of Engineering** was established in 1964, under the charter of the National Academy of Sciences, as a parallel organization of outstanding engineers. It is autonomous in its administration and in the selection of its members, sharing with the National Academy of Sciences the responsibility for advising the federal government. The National Academy of Engineering also sponsors engineering programs aimed at meeting national needs, encourages education and research, and recognizes the superior achievements of engineers. Dr. Charles M. Vest is president of the National Academy of Engineering.

The **Institute of Medicine** was established in 1970 by the National Academy of Sciences to secure the services of eminent members of appropriate professions in the examination of policy matters pertaining to the health of the public. The Institute acts under the responsibility given to the National Academy of Sciences by its congressional charter to be an adviser to the federal government and, upon its own initiative, to identify issues of medical care, research, and education. Dr. Harvey V. Fineberg is president of the Institute of Medicine.

The **National Research Council** was organized by the National Academy of Sciences in 1916 to associate the broad community of science and technology with the Academy's purposes of furthering knowledge and advising the federal government. Functioning in accordance with general policies determined by the Academy, the Council has become the principal operating agency of both the National Academy of Sciences and the National Academy of Engineering in providing services to the government, the public, and the scientific and engineering communities. The Council is administered jointly by both Academies and the Institute of Medicine. Dr. Ralph J. Cicerone and Dr. Charles M. Vest are chair and vice chair, respectively, of the National Research Council.

www.national-academies.org

The **Nigerian Academy of Science** is a private non-profit honorific organization of distinguished scholars in the pure and applied sciences. It was founded in 1977 and now has 126 members. Its mission is to promote the growth, acquisition, and dissemination of scientific knowledge, and to facilitate the use of science in the solution of problems of national interest. The academy is a voluntary organization, independent of government control, but collaborating with the government in the growth and development of science and technology in Nigeria. Professor David Okali is president the Nigerian Academy of Science. Professor Gabriel Ogunmola is former president.

COMMITTEE ON CREATION OF SCIENCE-BASED INDUSTRIES IN AFRICA

Rita Colwell (NAS), Chair
University of Maryland
College Park, MD

Bernard Amadei
University of Colorado
Boulder, CO

Nancy Bearg
Consultant
Washington, DC

Stephen Hoffman (IOM)
Sanaria, Inc.
Rockville, MD

Kung Shain-Dow
Consultant
College Park, MD

Winston Soboyejo
Princeton University
Princeton, NJ

NRC Staff

Michael Greene, Study Director
Policy and Global Affairs

Effie Bentsi-Adoteye
Policy and Global Affairs

Preface

In 1994 the World Bank approached the National Research Council (NRC) of the U.S. National Academies to propose a joint symposium on a novel topic. The Bank had identified what it called a "technology revolution" and an "economic revolution," which together were having a significant effect both on the ability of its client developing countries to compete in world markets and on the impact of the developing countries on the world economy itself. As Jean-François Rischard, vice president of the World Bank for finance and private sector development, explained:

> The world has entered a period of massive shifts in its economy. Among the many changes likely to occur, China will be the world's largest economy by 2020; digital television and telephone systems will completely change the way people and businesses communicate, and such traditional activities as deposit banking may become shadows of their former selves. Behind these changes and the reshaping of the world economy are two major forces: a technology revolution and an economic revolution.[1]

With these prescient words, he went on to describe the impact that information technology and its derivatives in transportation, communication, and software were having on business practices while creating new opportunities in the developing world. At the same time, he pointed out, 50 percent of the world's gross domestic product (GDP) and most of

[1]National Research Council and World Bank, *Marshaling Technology for Development*, Washington, DC: National Academy Press, 1995, p. 8.

its growth were being produced by countries that were not members of the Organisation for Economic Co-operation and Development (OECD), which at that time constituted all of the nonrich countries, by the World Bank's measure.

The theme of the symposium at which Rischard spoke was the central role played by knowledge in the new economy and the importance of access to and the creation of knowledge for all countries. At the conclusion of the symposium, the World Bank asked the National Research Council to develop a tool that the Bank could use in both assessing the state of practical knowledge in its client states and creating in these countries an awareness of the importance of applying that knowledge to exploit new opportunities and niches in the world economy. The Bank called the putative methodology a "knowledge assessment."

The moment was opportune for the NRC to take on this assignment. It had just completed a three-year project in Indonesia, funded by the World Bank, devoted to applying science and technology to industrial development. As part of the U.S. National Academies complex, the NRC is able to recruit and convene experts on nearly every technical topic and every technology from around the world, and many highly distinguished scientists and industrialists were brought to Indonesia during that period. In the process, the Bank and the NRC learned an important lesson: the obstacles to technology development are most often not themselves technical, but rather social, economic, and political. They also found that the local business and scientific communities were quite well informed about the problems and obstacles within their own countries, and thus the combination of the international and local experts working together produced a powerful synergy that aided in finding solutions. At the same time, this synergy created a constituency for change within the country, centered around the local participants themselves.

Out of this experience arose the interactive version of the knowledge assessment, in which a small group of international experts plays the role of venture capital investors visiting the country to explore the potential for investment. They meet with different groups of experts representing agriculture, labor, industry, the scientific community, education, and finance to evaluate the potential for investment in new technologies. During that meeting, they select a small number of diverse technologies that, although successful elsewhere, are not presently being exploited within the country's economy, and they proceed to explore the obstacles that might exist. The next step is a series of workshops called hypothetical case studies, each based on one of the selected technologies. International business experts and engineers, often the CEOs of successful companies based on the technology, work with local experts to create a business plan for a hypothetical enterprise based on the technology in question. In the process, they expose the problems in the local environment, identify the conditions for creating a profit-making enterprise, and assemble cost estimates. The result of the knowledge assessment, after combining the conclusions of the different hypothetical case studies, is a series of recommendations to donors, government, the private sector, and other local organizations for making the country more hospitable to new technologies and science-based enterprises. In recent years, this methodology was tested in Panama and El Salvador, and the pilot study took place in Prince Edward Island, Canada.[2]

The Nigerian study described in this report seemed to be a useful, if not ideal, application of the knowledge assessment methodology. The technologies selected had not yet appeared in practice in the country, and the knowledge assessment could be used to identify the obstacles and recommend actions to encourage enterprises to put them to use. There was one difference, however. The original knowledge assessment was intended to analyze the economy, and the case studies were merely vehicles to study different sectors. In the Nigerian study, it was the intention that the enterprises actually come into existence.

A feature of the knowledge assessment method is that the workshops between the international experts and the local business and scientific community serve as the exclusive source of all the information generated—that is, the participants provide all information used in the study, even though, in reality, some of it may be misinformed or even wishful thinking. The resulting report can serve as a useful guide to the potential investor or entrepreneur, but with the caveat that the information presented must be verified at the time and specific location it is meant to be applied. For this reason, one single technological solution is not presented; rather a set of choices—amorphous versus crystalline silicon photo cells, ultraviolet versus ceramic water filtration, and several different artemisinin combination malaria therapies—are offered in every case. In fact, in a fast-moving field like artemisinin-based malaria therapy in which an international consortium is striving to support a global subsidy, the committee found it difficult to remain up-to-date, even during the review and publishing process.[3] Thus this report may not offer concrete

[2]This process is described more fully in the following reports: National Research Council, Committee on Knowledge Assessment, *National Research Council Prospectus for National Knowledge Assessment*, Washington, DC: National Academy Press, 1996; and National Research Council, Committee on Knowledge Assessment, *Lighting the Way: Knowledge Assessment in Prince Edward Island*, Washington, DC: National Academy Press, 1999.

[3]For example, in the spring of 2007 it was announced that the pharmaceutical giant Sanofi-Aventis had reached an agreement with the Drugs for Neglected Diseases initiative (DNDi) and Médecins sans Frontières to offer a compound drug meeting the price and quality standards of the World Health Organization at less than $1 a course. Another low-cost

and certain answers, but it can be used to suggest the questions that need to be asked. As for the usefulness of the report to the Nigerian government, the fact that prominent local businessmen, businesswomen, and scientists believe in the truth of its statements—for example, those related to education, infrastructure, and regulation—is itself important information about business confidence.

ACKNOWLEDGMENTS

The committee is grateful to the Nigerian Academy of Science and its president at the time of this study, Professor Gabriel Ogunmola, for their help and collaboration in carrying out this work. The efforts of all the Academy members and other experts named in the appendixes who voluntarily participated in the workshops and are primarily responsible for the information that appears in the report are sincerely appreciated. A strong expression of gratitude is extended to Hellen Gelband of the Institute of Medicine. Ms. Gelband was project director of the study *Saving Lives, Buying Time,* which proposed the global subsidy for artemisinin-based malaria drugs, and her guidance through that thicket was invaluable. Krishna Kumar of Howard University provided the cost estimate for artemisinin combination therapy manufacture. The report was edited, meticulously as always, by Sabra Bissette Ledent.

This report has been reviewed in draft form by individuals chosen for their diverse perspectives and technical expertise, in accordance with procedures approved by the NRC's Report Review Committee. The purpose of this independent review is to provide candid and critical comments that will assist the institution in making its published report as sound as possible and to ensure that the report meets institutional standards for objectivity, evidence, and responsiveness to the study charge. The review comments and draft manuscript remain confidential to protect the integrity of the deliberative process.

We wish to thank the following individuals for their review of this report: Dana G. Dalrymple, U.S. Agency for International Development; Thomas L. Dixon, TechnoServe (Tanzania); Barrett Hazeltine, Brown University; Uford Inyang, National Institute for Pharmaceutical Research and Development (Nigeria); Akinlawon L. Mabogunje, Nigerian Academy of Science; and R. Rhodes Trussell, Trussell Technologies Inc.

Although these reviewers provided many constructive comments and suggestions, they were not asked to endorse the conclusions or recom-

compound is in the process of registration in Brazil. However, because this report does not identify which compound should be produced in Nigeria, these developments do not alter its conclusions.

mendations, nor did they see the final draft of the report before its release. The review of this report was overseen by Queta Bond of Burroughs Wellcome Foundation. Appointed by the National Research Council, she was responsible for making certain that an independent examination of this report was carried out in accordance with institutional procedures and that all review comments were carefully considered. Responsibility for the final content of this report rests entirely with the authoring committee and the institution.

Contents

Summary 1

1 Introduction 11
Nigeria, 13
Filling the Gaps: Three Technologies to Meet Three Major Needs, 15
Doing Business with the Poor, 19

2 Methodology 26
Hypothetical Case Studies, 28
Nigerian Adaptation, 29

3 The Case Studies 31
Solar Photovoltaics, 31
Financing the Enterprise, 36
Water Purification, 37
Antimalarial Artemisinin Combination Therapy (ACT), 45

4 Meeting Needs 55
Mobilizing the Private Sector to Provide Public Goods, 57
The Special Case of Artemisinin, 59
Linking Science-Based Enterprises to Research, 61

5 Recommendations 63
Incentives for Private Companies to Provide Public Goods and Services, 63
The Challenge of Artemisinin Combination Therapies, 66
Consumer Education and Training, 67
The Role of Philanthropic Foundations and Donor Agencies, 68
The Role of the Nigerian Academy of Science, 70

Appendixes: Workshop Reports 75

A Solar Photovoltaics: Hypothetical Case Study 77
B Water Purification: Hypothetical Case Study 90
C Artemisinin-Based Malaria Therapy: Hypothetical Case Study 108

Summary

This report is the product of collaboration between the U.S. National Academies and the Nigerian Academy of Science. Nigeria is an example of a mostly rural developing country whose government is unable to provide some basic services, such as potable piped water and electric power, to a large proportion of the population in an affordable manner. In other developing countries, many of these services are available from private, profit-making companies using readily accessible technologies. This report examines how Nigeria can mobilize private companies to provide some basic services that might be sustainable and cost-effective for government, company, and consumer.

In Nigeria, about two-thirds of the population lacks safe water and access to the electricity grid. A similar proportion lacks effective treatment for malaria, a major cause of child mortality and loss of productivity. Because the government does not provide malaria drugs (or other medicines as well), most people buy their medicines privately. However, in Africa the malaria parasite has become immune to the existing low-cost drugs, and, as for HIV/AIDS, the newer, more effective treatment is too expensive for the majority of patients. In response to this problem, the international community, including the World Health Organization (WHO) and World Bank, has been exploring ways to subsidize these drugs, which, if successful, will expand manufacturing opportunities for the new products.

The word *sustainable* is generally applied in the international development context to solutions that do not depend on donor funds or ongoing government financial support. Services provided by private enterprises may be considered sustainable when the enterprises are able to make a profit. People without electrical power, safe water, and effective medicines are usually poor, but any firms that provide the poor with these essential services must be able to profit from doing so. Thus, they would require a business model designed for serving a large number of clients who have very little disposable income. The extremely dense urban environment and highly dispersed rural communities that characterize the bottom of the economic pyramid in the developing world require a new approach. Some companies have developed such models and do relatively well in other countries. Elements of these business models include the following:

- a focus on the price performance of products and markets
- incorporation of innovative hybrid solutions that use advanced technologies blended with the existing culture and with products designed to work in hostile environments characterized by, among other things, an irregular power supply, contaminated water, low skill levels, and unreliable infrastructure
- an emphasis on reducing, conserving, and recycling resources, especially packaging
- adoption of innovative processes for local manufacture
- application of innovative methods of financing, distribution, and marketing

Microcredit, service contracts, and franchising opportunities also are important elements of the business models.

As limiting as the conditions in developing countries seem to be, the great advantage is the huge number of potential clients. An estimated 100 million Nigerians lack safe water, electric power in the home, and effective malaria therapy, or more than the total populations of all but a handful of countries. In India and other big countries with large numbers of poor people, companies (including multinationals) aiming at the customer base at the wide bottom of the economic pyramid have produced new, innovative products and services at substantial profit to themselves as well as with benefits for their customers.

This study aims to demonstrate that for the three examples chosen—solar electric power, safe household water, and effective malaria therapy—it should be possible to make a profit providing these products in Nigeria without direct government support (although for malaria drugs, a global subsidy of some kind probably would be needed). Nevertheless, actions the government might take to encourage private sector participation and extend the benefits to large segments of the population are described in this report.

METHODOLOGY

The methodology used to demonstrate the viability of the business models is called a hypothetical case study. It was originally devised as part of the knowledge assessment methodology prepared by the U.S. National Academies for the World Bank to identify opportunities for developing countries to find niches in global markets by exploiting technologies not yet in use in the countries.

A workshop was held on each of the selected technologies (solar energy, December 8–9, 2005, in Lagos; safe water, December 12–13, 2005, in Lagos; malaria therapy, April 24–25, 2006, in Paris). The workshops were designed to exploit the interactions among international experts and entrepreneurs who had successfully created enterprises based on each technology in similar countries and local scientists and business experts who understood the economic and social environment of Nigeria. The aim of all the participants was to identify the conditions for success and produce a sample business plan and cost estimate for creating an enterprise that would exploit the technology in Nigeria. Nearly all the information used in the report was supplied by the expert and local participants in the workshops. For this reason, the emphasis is not on providing formulas and specifications, which must be checked and updated, but on providing a range of technological choices and the questions that potential entrepreneurs and investors must ask in selecting among those choices. The workshops identified obstacles and proposed solutions to them. When these solutions require action by government, private sector associations, and other institutions, the actions are reflected in the report's recommendations, which are summarized here and presented in detail in Chapter 5.

THE TECHNOLOGIES

Solar photovoltaic systems are installed in a home or community. The example used by the workshop was the Solar Electric Light Company (SELCO) in Karnataka State, India. For the time being, the solar cells would have to be imported into Nigeria, but small local companies can provide installation and maintenance services. In the SELCO model, maintenance services are very important, and they must be provided onsite and regularly. With the proper consumer credit, the system can be made affordable to poor homeowners and small rural businesses. Recent rises in the cost of the kerosene presently used for lighting in rural villages may further motivate Nigerian companies to install and service solar home systems, especially because government support of the local manufacture of solar photovoltaic units is already becoming available.

Safe water can be provided in many ways. The workshop considered both ceramic filters, which can be operated in the home and were designed by Potters for Peace in Nicaragua, and the more elaborate ultraviolet (UV) filters, made by WaterHealth International of California and distributed in several developing countries. The UV filters can be purchased by a community or by franchisees who sell safe water to households. Because both filtering units must be operated by the user, training is an important part of the service. Nevertheless, the price of the water to the consumer is low, the savings on the medical services avoided from water-borne diseases is high, and most families should be able to afford such technologies without assistance.

Although they are now recommended by WHO as the first-line treatment for uncomplicated malaria, artemisinin combination therapies are used in only a minority of cases, largely because countries cannot afford to purchase these high-price medicines in sufficient quantities, even though most have adopted them as official policy. Nigerian companies may have an early opportunity to become part of the global value chain, while assuring reliable supplies to Nigerians. But the viability of this model depends on the international community agreeing on a way in which it can reduce the cost of these drugs to consumers. The current situation—an acknowledged crisis—is the subject of deliberations that should lead to some resolution relatively soon. Nigerian researchers growing the plant from which artemisinin is extracted (*Artemisia annua*) and officers of Nigerian pharmaceutical companies producing malaria medicines participated in the workshop.

RECOMMENDATIONS

Incentives for Private Companies to Provide Public Goods and Services

Nearly two-thirds of the Nigerian population does not have home lighting, safe drinking water, or effective malaria therapy. Lacking the resources to more than double the number of people served by the electricity grid and piped water network, the Nigerian government can turn to the highly active private sector to provide these basic products and services through a program of financial incentives and technical assistance to encourage responsible service. If a large proportion of poor and isolated communities are to be served, a solution engaging the private sector may require the involvement of a large number of new companies, many of them start-ups. First-stage financing or venture capital will perhaps be required for the majority of them.

Recommendation: **The Nigerian government should develop a system of incentives to encourage private companies to sell and service solar electric systems for the home to rural residents who are not connected to the national grid. The Nigerian government also should develop a system of incentives to encourage private companies to sell and service water purification systems to communities that are not served by municipal or national water supplies, or to produce household filtering systems for safe drinking water.**

Consumers of home solar electric systems will need consumer credit to purchase their units. Financial institutions should be encouraged and trained to make small loans for such systems, including service contracts, ultimately secured only by the system itself, as has been done in other countries.

Recommendation: **The Nigerian government should work with banks and other financial institutions to establish microloan funds that would be dedicated to providing consumer credit for home solar electric and water filtration systems in rural areas.**

Many householders are not aware of the advantages to their families and the improvements in their lifestyles and educational and business opportunities of having electric lighting, radios and televisions, and refrigerators in their homes. Many also do not realize the harmful effects that contaminated water can have on their children and the cost to themselves and the nation.

Recommendation: **The Nigerian government should sponsor an educational campaign to encourage people to invest in electric power and safe water.**

Existing legislation might actually make it legally hazardous for private companies to bring electric power into people's homes, because this service is now the exclusive mandate of a federal ministry.

Recommendation: **A legal remedy should be found to the prohibition on private companies or individuals providing power to homes in Nigeria. In the context of legislation on renewable energy sources, it might give additional impetus to the proposed program.**

The Challenge of Artemisinin Combination Therapies

The only first-line therapies for uncomplicated falciparum malaria (the most deadly form of the disease) recommended by the World Health Organization are artemisinin combination therapies (ACTs), based on extracts of *Artemisia annua*, a plant commonly found in the wild. The production of ACTs by Nigerian companies is unlikely to succeed without various types of assistance by the government. A viable market, not just in Nigeria but globally, depends on reducing the price of ACTs to consumers, which will require subsidization, even assuming efficient manufacturing and small profit margins.[1] An international effort is under way to provide these subsidies in every country.

Meanwhile, Nigerian pharmaceutical companies and public health officials may not be sufficiently informed of developments in the international arena that will affect their plans, including progress toward subsidization and the global technical standards for manufacturing ACTs. The government could stimulate ACTs production by ensuring that all interested officials and companies along the supply chain, from cultivation of *A. annua* through manufacture and packaging of ACTs, are informed of current and developing conditions.

Recommendations:

The Nigerian government should support private sector participation in the global ACTs market. It should do this by tracking international developments related to the economic and technical requirements of all aspects of ACTs production and establishing formal and informal links to academic, public, and for-profit entities that could play a role in that production. The Nigerian Academy of Science could play an important role by convening potential participants for the exchange of information with the government and the international community.

The Nigerian government should ban all counterfeit drugs, illegal clones, low-quality products, and artemisinin monotherapies from the Nigerian market by means of proactive enforcement against illegal activity. It also should prohibit the advertising of such products.

Duties and other impediments to the importation of the equip-

[1]The main proposal under development by the global malaria community was introduced in the 2004 Institute of Medicine (IOM) report *Saving Lives, Buying Time: Economics of Malaria Drugs in an Age of Resistance*, Washington, DC: National Academies Press, 2004.

ment, raw materials, solvents, and other materials needed for ACTs research and production should be removed. The Nigerian government should support training and research on the agronomy and selection of the best cultivars of *Artemisia annua* in collaboration with the global malaria community. Public health laboratories should participate in surveillance to determine the levels of resistance to possible partner drugs for different ACTs formulations.

Consumer Education and Training

In every modern society, consumers are confronted by a bewildering array of products and services that affect their health, well-being, and economic security. However, most countries offer little formal consumer training, and so consumers are expected to educate themselves by means of the media, their friends and community, and commercial advertising. Sometimes, consumers receive advice from the public regulatory agencies or private nongovernmental organizations (NGOs) that monitor product safety, drug efficacy, and truth in packaging. When these watchdog agencies are ineffective or nonexistent, or when populations are illiterate or isolated, public agencies must step in and provide more explicit consumer education through the schools or in public campaigns or pronouncements.

Recommendation: **The high mortality rates in Nigeria from diarrheal disease and malaria argue that the Nigerian government should offer health education and training in the schools that would address the importance of safe drinking water, how to maintain a sanitary water supply, and how to choose effective medicines. Other relevant topics might include nutrition, hygiene, safe sex, and the prevention and treatment of common diseases such as respiratory infections, diarrhea, and HIV/AIDS. This program should be supplemented by public pronouncements on topics such as filtering water and a public information campaign on the importance of using the most effective antimalarial therapy—currently ACTs—and the necessity of completing the course of treatment.**

The Role of Philanthropic Foundations and Donor Agencies

The proposals offered here for government suggest a new approach for philanthropic foundations and donor agencies. Ideally, any donor should have two objectives. First, enable the target community to find and implement solutions to its own problems rather than giving it a particular solution—that is, as the popular saying goes, teach people to fish instead of giving them fish. Second, create sustainable solutions that do not require an ongoing supply of donor funds.

Development philanthropy should be considered a kind of investment in essential goods and services. The goal should be to create *enterprises* that will work to continue to provide the goods and services that represent sustainable solutions to socioeconomic problems.

For donor agencies, this path is a more difficult one, because the creation of successful enterprises is less well understood in the philanthropic community than grant giving, and a natural failure rate characterizes even the most fertile investment plans. The effort must be spread among many enterprises to improve the probability of success; it will resemble a balanced investment portfolio more than a philanthropic grant program.

Some of the NGOs that often serve as agents of the donor agencies should also modify their methods of operation. They are adept at organizing demonstration projects, such as one that might illustrate the feasibility of installing solar photovoltaic systems in villages by donating some and teaching the recipients to use them. But experience has shown that, relieved of the responsibility of selecting and paying for the unit, the recipient of such a demonstration has little motivation to maintain it or use it effectively, and the demonstration project often ends up showing villagers that such utilities need not be paid for and have a limited lifetime. This result is detrimental to entrepreneurs who seek to sell and service such units, and it should not be part of the proposed approach.

Recommendation: **Philanthropic foundations and donor agencies should orient some of their activities in developing countries toward creating and supporting profit-making enterprises that would provide public-benefit goods and services to poor people. Grants should be replaced in spirit with first-stage financing or investments, and the portfolio should be broad enough in diversity of enterprises with different business plans and different technologies to raise the probability of financial success in this area, where there is relatively little experience.**

The Role of the Nigerian Academy of Science

The Nigerian Academy of Science is positioned to strengthen its public role and to help further the goal of sustainable development.

Recommendations:

The Nigerian Academy of Science should establish a program similar to the U.S. National Academies' Government-University-Industry Research Roundtable (GUIRR), focusing on scientific and techno-

logical issues of common concern. It should sponsor and convene workshops of experts drawn from the government, academia, and industry, and the reports of the workshops should be published by the Academy.

One early topic for a meeting of the Nigerian roundtable and a workshop of experts should be the production of ACTs in Nigeria. Invitees would include agriculturalists, pharmaceutical companies, and government health officials. The workshop would seek to help interested companies become informed about international efforts to subsidize the price of ACTs and the requirements for cGMPs (current Good Manufacturing Practices) certification and WHO prequalification to manufacture ACTs.

Another, equally important early topic for the Nigerian roundtable should be the importance of safe potable water to public health in Nigeria. A follow-on workshop that includes experts from academia, government, NGOs, and the private sector would bring the issue to the public. Especially useful participants would be representatives of the Ministry of Health, the Federal Institute of Industrial Research, the International Center for Business Research, the National Agency for Science and Engineering Infrastructure, and the Nigerian Association of Small and Medium Enterprises. Importantly, such a workshop could urge the government both to clarify the law on the right to provide potable water to households and to mount a campaign in favor of filtered purified water to combat diarrheal disease.

The U.S. National Academies should be prepared to assist the Nigerian Academy of Science to organize the Nigerian roundtable, if requested, in view of the National Academies' long experience with GUIRR. The National Academies also might assist by arranging for international experts to participate in Nigerian Academy–sponsored workshops dealing with solar energy, safe water, and ACTs.

U.S. scientific agencies with international programs, such as the Office of International Science and Engineering of the National Science Foundation and the Fogarty Center of the National Institutes of Health, should guide the exchange programs between the United States and countries such as Nigeria toward cooperation in helping small and medium enterprises to provide public goods and services. Many U.S. scientists have valuable expertise in linking research to enterprise creation.

1

Introduction

This report is the product of collaboration between the U.S. National Academies and the Nigerian Academy of Science. The subject of the collaboration was a study of the ways in which science-based private enterprises might be created and promoted in Nigeria and other developing countries in order to provide science-based products and services that government is unable to supply in a timely and sustainable manner. Examples of these services are electric power and safe household water in rural areas. In other developing countries, lack of safe water and lack of home or small business electric lighting are problems that have generated entrepreneurial solutions through readily accessible technologies. But in Nigeria private companies have generally not been viewed as an instrument of government policy to extend basic services to the underserved.

Malaria presents a similar situation. This devastating disease, which kills one million people worldwide every year, must be tackled with new drug treatments to replace those that have lost their effectiveness both to cure the disease in individuals and to reduce its spread. The local private sector may be able to join government and donors in seeking a solution.

The following statement of task was presented to the committee:

> In collaboration with the Nigerian Academy of Science, an ad hoc committee will (1) develop a model for dissemination of technologies of social benefit and creation of science-based enterprises in Nigeria; (2) select three technologies related to health, agriculture, and small-scale

industry with potential for commercialization in Nigeria and Africa; and (3) carry out a Knowledge Assessment of the selected technologies to identify opportunities and barriers to creating the science-based enterprises in Nigeria. This approach involves active participation by the local business community, local and national commercial banks, producers, and scientists and engineers, interacting with international experts, to explore the prospects for enterprises based on the selected technologies. The report will recommend actions by government, the private sector, and the national academies of Nigeria and the United States to encourage science-based enterprises.

The three technologies chosen by the committee were solar photovoltaics, water purification, and malaria therapy. The methodology selected, a knowledge assessment (described in Chapter 2), was used by the committee to illustrate ways in which the technologies could be profitably applied in Nigeria. The committee proposed a development model in which the private sector would be able to provide the technology-based products and services at a profit. Such a model offers the possibility that, after a period of incentive and encouragement, government or donor support would not be required in the future, thereby satisfying most definitions of economic sustainability.

Nigeria is an appropriate test bed for an approach that combines government and donor support and resources to enable the private sector to manufacture and provide science-based solutions to basic needs problems. As an oil exporter with a positive foreign exchange balance, Nigeria has a source of funds that could be employed to test the hypothesis. Nigeria also has several excellent universities, and the Nigerian Academy of Science is populated by many world-class scientists. In fact, Nigeria is famous worldwide for its entrepreneurial class, which includes modern manufacturing and extractive industries.

And yet despite these and other assets, Nigeria remains in the World Bank's low-income category, and 60 percent of the population lives below the poverty line; life expectancy is 46 years.[1] About two-thirds of Nigerians have no access to the electricity grid or safe water. A similar proportion of the people are at risk of malaria, which is a major cause of child

[1]For a listing of countries in the various income categories, see World Bank, http://siteresources.worldbank.org/DATASTATISTICS/Resources/CLASS.XLS. The poverty line is defined by the "FEI method," which is based on annual expenditures on food compared with the cost of the minimum daily allowances recommended by the World Health Organization. See Ben E. Aigbokhan, "Poverty, Growth and Inequity in Nigeria: A Case Study," AERC Research Paper 102, African Economic Research Consortium, Nairobi, November 2000, http://www.aercafrica.org/documents/rp102.pdf. For life expectancy, see the U.S. Central Intelligence Agency's *World Factbook*, https://www.cia.gov/cia/publications/factbook/geos/ni.html.

mortality and loss of economic productivity. In Africa, the malaria parasite is becoming resistant to most low-cost and readily available drugs, and a newer effective treatment is currently too expensive for the majority of patients.

For each of these fundamental problems holding Nigeria back and imposing suffering on its people and economy, the technology that could solve the problem is available for transfer and incorporation into the private sector. The technologies for water purification and solar electric lighting have been applied successfully elsewhere in the developing world, including in Africa, and the current most effective malaria drug treatment, artemisinin combination therapy (ACT), could be produced by the Nigerian private sector.

These three technologies explored in the study—solar photovoltaics, water purification, and effective malaria therapy—and the associated business models were selected by the two science academies to serve as case studies in order to demonstrate how the government-sponsored participation of private sector enterprises might be used to provide basic services. A workshop was conducted for each technology. At each workshop, several foreign businesspeople who had successfully exploited the particular technology to create profitable enterprises in other developing countries were brought together to collaborate with a diverse group of Nigerian businesspeople, scientists, financial experts, and others. Together, they designed a business plan for a hypothetical Nigerian enterprise to produce the technology, drawing on the foreign experience while taking into account the social, economic, and cultural environments of Nigeria. The workshop reports produced, one for each technology, appear in the appendixes to this volume. They constitute the "data" used to evaluate the hypothesis that such enterprises could be successful and effective in Nigeria, and the workshop results underlie the steps recommended in Chapter 5 for exploiting these technologies for providing basic services.

NIGERIA

In March 2006, Nigeria carried out its first census in 15 years. The size of the country's population has been a matter of controversy, and several states did not accept the results of the last census in 1991. Estimates of the population range from 135 million to 162 million.[2] However, there is one figure that appears to be supported by consensus, if not census, and it is a staple of everyday conversation, with little variation. In Nigeria, 100 million people have no access to safe water; 100 million have no access to

[2]The figure 135 million is found in the *World Factbook*, ibid. The figure 162 million is found at http://www.internetworldstats.com/africa.htm#ng.

the electricity grid; and malaria strikes 100 million people each year. In fact, these numbers are generally supported by the responsible agencies. United Nations (UN) agencies report that only between 50 and 60 percent of households have access to improved drinking water.[3] The Global Energy Network Institute (GENI) reports that few rural households are electrified.[4] And the Roll Back Malaria program of the World Health Organization (WHO) suggests that the 100 million figure for malaria incidence in Nigeria, where 80 percent of the population is exposed to malaria and 60 million experience more than one attack a year, might be low.[5] This study is dedicated to the "100 million," which is its target population.

Nigeria is the most populous country, and one of the more advanced, in Africa. As an oil exporter, it has exchange reserves of $28 billion. Rising oil prices have brought new money, some of which will be dedicated to science. The government announced plans in June 2006 for a $5 billion endowment fund for science and technology, drawn from the oil revenues and supplemented by donors and the private sector. An independent Nigerian National Science Foundation will be created to disburse research grants, establish new scientific universities, and equip existing research groups.[6]

The gross domestic product (GDP) of Nigeria is $83.36 billion, or $1,500 per capita, with a real growth rate of 5.3 percent. The economy consists of about 53 percent industry, dominated by oil production; 17 percent agriculture; and 30 percent services. Oil production is 2.5 million barrels per day, of which about 2 million barrels per day are exported. In 2004 electricity generation was 19 billion kilowatt-hours, of which 20 million kilowatt-hours were exported to neighboring countries.

Safe water and malaria control are key elements of the United Nations' Millennium Development Goals, for which Nigeria has pledged universal coverage by 2015. However, since 1990 indicators of the rate of progress have fallen.

On Transparency International's Corruption Perception Index (CPI), Nigeria is rated 1.9, ranking 152 out of 159 countries. The CPI score is based on business people's and country analysts' perceptions of a country's degree of corruption. It ranges between 10 (highly clean) and

---

[3]UNICEF, http://www.unicef.org/infobycountry/nigeria_28236.html; Millennium Development Goal Indicators, http://unstats.un.org/unsd/mdg/Data.aspx?cr=566.

[4]Global Energy Network Institute, http://www.geni.org/globalenergy/library/media_coverage/africa-renewal/energy-key-to-africas-prosperity.shtml.

[5]Communication Initiative, http://www.comminit.com/trends/issuestrends/sld-2098.html.

[6]"A Foundation for Africa," *Nature*, August 3, 2006, 442, 486, http://www.nature.com/nature/journal/v442/n7102/index.html#nf.

0 (highly corrupt).[7] Nigeria also appears on the 2006 World Bank list of poorly governed "fragile states" that are at risk of terrorism.[8]

In 2005 the Nigerian government prepared a National Economic Empowerment and Development Strategy (NEEDS) in cooperation with the International Monetary Fund (IMF).[9] It states, in part:

> Nigeria has become a nation of traders, with a very weak and stagnant domestic private sector. Other frequently cited problems in efforts at growing the private sector include the poor state of physical infrastructure; the high cost and limited access to appropriate financing; insufficient domestic demand and the low level of patronage by public sector institutions; the high cost of imported raw materials, equipment, and spare parts; and the lack of skilled labour. Growing the private sector also hinges crucially on domestic policies, environmental factors, and investment flows.

One of the solutions suggested is to

> redefine the role of government as a facilitator and promoter in the economy, recognizing that market failures in developing economies require targeted incentives and interventions in specific areas to promote specific sectors and industries. The government hopes to complement the usual enabling environment model of development with some targeted entrepreneurial interventions to bolster weak and vulnerable sectors.

FILLING THE GAPS: THREE TECHNOLOGIES TO MEET THREE MAJOR NEEDS

Although Nigeria is an industrial leader in West Africa, with an educated and entrepreneurial population, foreign currency reserves in the bank, and currently a functioning democracy, 100 million of its people do not have electric power, 100 million lack safe water, and 100 million suffer from malaria without effective treatment. In part, the solutions to these problems are related to infrastructure, and for water and electric power the solutions could be found in expanding the electricity grid and build-

---

[7]Transparency International, http://www.transparency.org/policy_research/surveys_indices/cpi/2005/.

[8]World Bank, "Fragile States: The Low Income Countries Under Stress (LICUS) Initiative," http://web.worldbank.org/WBSITE/EXTERNAL/NEWS/0,,contentMDK:20127382~menuPK: 34480~pagePK:34370~piPK:116742~theSitePK:4607,00.html.

[9]International Monetary Fund, "Nigeria: Poverty Reduction Strategy Paper—National Economic Empowerment and Development Strategy/International Monetary Fund," IMF Country Report No. 05/433, Washington, DC, 2005, pp. 48–52, http://www.imf.org/external/pubs/ft/scr/2005/cr05433.pdf.

ing municipal water facilities (Nigeria is not deficient in groundwater). However, like many developing countries Nigeria is a rural country, and expanding the electricity grid and building municipal water processing plants to serve every site may not be cost-efficient or a budget priority.

Household Water and Electrical Technologies

Decentralized energy and water supply facilities represent unique opportunities for remote rural developing communities to provide the basic necessities of life to those in need at lower cost. Small, stand-alone systems are more appropriate and sustainable in remote areas. Such systems not only are more cost-effective for the user, but also are more flexible, less disruptive to communities, and more likely to meet people's needs, improve the future of communities, and act in harmony with sustainability principles. Decentralized systems are often the only realistic option for providing developing communities with water and energy.

And yet stand-alone systems do have limitations. Their success may depend on market incentives for companies, investors, and consumers and sustainable quality control and maintenance. Moreover, because they are beyond government control, they may be of lower priority for governments.

Solar photovoltaic systems and water purifiers, the technologies considered in this report, fall at the low-cost end of the spectrum of regional electric power and safe water supplies. The criteria for their selection include the following:

- They fulfill a basic need of the people that is not being met by government for a majority of the population.
- They can be used effectively to provide services that fill the need at the village or household level.
- They can be provided by private enterprise without direct government intervention.
- They can be installed and maintained locally at a profit for the provider, and thus they have the potential to be sustainable indefinitely.

Neither solar photovoltaic systems nor water purifiers involve new technology or custom-designed units. Both can be adopted from currently available models and constructed from ready-made items, although not all of the items are necessarily available today in Nigeria. The effectiveness of the technologies in the village environment has been demonstrated, at least for short periods, by donor- and government-supported projects many times over decades.

Sustainability, however, is another question. In the great majority of projects worldwide, the units were donated to villagers. Unit maintenance was then either neglected or included in the cost to the donor. But a characteristic of all donor-supported projects and many government ones is that eventually the funding for the project ends, or funds are diverted to other worthy uses after successful demonstration of the technology. Within a short time, the solar or water purification units fall into disuse or disrepair without service, and yet the need is as great as before. The missing element in sustainability is real ownership of the technology by villagers. Because they were given the technology as a handout, they are less likely to use and maintain it than if they had selected and bought it with their own hard-earned money, which a private sector approach would require.

As for the government's role, electric power and drinking water are typically in the government realm in most countries, including Nigeria, unlike some other basic needs. For example, food is not supplied by the government, but it is available to most citizens through a complex web of farm supply, farmers, distributors, and markets that manages to put bread or cassava on nearly every table. Clothing is available everywhere through a combination of mass production, cottage industry, and low-cost imports with effectively no government regulation or participation. Housing in Nigeria and most other countries is largely in the hands of the private sector, formal and informal, although many governments step in when the poor experience serious shortages. In the health care realm, the generally large pharmaceutical companies provide those who can afford them with common medicines through local distribution networks and markets. In each of these areas, the needs are met, continually and sustainably, by the private sector; people are employed, and profits are made. The question then is: could the government and donors find a way to stimulate the private sector to also provide small-scale household electric power and safe water as an alternative to extending the national electricity grid or building municipal or national water facilities?

Malaria Treatments

Malaria control and treatment present a somewhat different problem. Malaria drug therapy is widely available at a low price in Nigeria, and it is frequently featured in billboard advertising. However, the common low-cost medicines such as chloroquine and Fansidar are largely ineffective because the malaria parasite is becoming resistant to them. In fact, the only known therapy for which no known resistance has appeared is artemisinin, derived from *Artemisia annua,* an herb endemic to China that has been adapted to cultivation in parts of Africa. Drugs containing artemisinin or its derivatives are available in Nigeria, but at a cost more than 20 times higher than that of the common low-cost remedies. It is, nevertheless, cost-effective because malaria can be fatal and artemisinin cures the disease (ineffective medicines are not cost-effective at any price). Indeed, WHO has recommended it as the sole approved remedy for malaria. And yet the cost puts it beyond the means of the majority of sufferers, which leaves it up to the government to find a way to supply this treatment to all patients at an affordable price.

A solution may be within reach, however, because of the role proposed for the international community by WHO. The cost to the government of providing free artemisinin-based drugs to all malaria sufferers in Nigeria at current prices would be more than $200 million a year. The prescribed artemisinin-based drug, in a formulation called artemisinin combination therapy, is available locally at a price exceeding $20 per course of treatment, whereas chloroquine can be purchased for $1 per course of treatment, a high price for an ineffective drug but one that is affordable for everyone. An international movement organized by WHO and involving several other organizations is seeking to subsidize the price of ACTs for patients, so that ACTs will be accessible to most people everywhere. However, the only formulations that will qualify for the subsidy are those manufactured according to WHO's standards. At present, only one drug company, Novartis of Switzerland, meets the standards. Its product, Coartem, is available in Nigerian pharmacies at more than $20 per course of treatment.

Nigeria has the capability to produce ACTs, and several are locally available, although at high prices. But these products will not be competitive when and if the subsidy goes into effect, unless the manufacturers meet WHO's manufacturing standards as mandated by the subsidy. At this time none do, but several local pharmaceutical companies claim to soon be able to meet WHO's manufacturing standards, which would allow them to receive the subsidy and sell ACTs they would manufacture in Nigeria, as well as export them for sale at the subsidized price.

The Nigerian government has announced a program to provide ACTs free to children under five, which would be a great boost to health and child survival. It has also declared ACTs the only approved malaria treatment. The nation would gain economically through higher productivity and lower workplace absenteeism if ACTs were provided at a low price to all. Local production of effective malaria medicines might enable realization of these gains, whether or not the subsidy materializes. If the worldwide subsidy is applied, local manufacture will not be required because the medicine can be imported at the subsidized price. But, if local manufacturers were prequalified, they could export ACTs regionally at the subsidized price as well as serve the local market.

DOING BUSINESS WITH THE POOR

Access to a commercial good or a sustainable service generally depends on the provider of the good or service being able to profit from the enterprise. Donated products will always depend on the existence of donated funds, funded research programs, or local initiatives that, by their very nature, are short term. Because people without electricity, safe water, and effective antimalarial medicines are primarily poor, it is critical to test the proposition that a firm can profit by providing essential goods and services to the poor. What, then, is a distinct but workable business model for serving those with no disposable income?

The usual philanthropic route to providing goods and services such as water, electric power, and medicine to the poor is to engage nongovernmental organizations (NGOs) to provide the products without cost to the recipient community. But decades of experience show that items received at no cost are considered of no value and often are not used, maintained, or replaced when needed. Similarly, in a variation on the "tragedy of the commons," devices donated to a village or community are considered to belong to everyone, and therefore to no one in particular. Thus, no one is willing to maintain or replace them when needed.

By contrast, companies, large and small, are profiting by serving—and charging—the poor. Some of the business models employed, though not following a common pattern, have features that distinguish them from those of normal free-market companies. Some of these models, such as combining banking services with product sales, may simply not be permitted in the United States and other industrialized country economies. Some of the best known of these models are reviewed in this chapter and then examined later in detail in this report.

The subject of businesses that serve the poor has been studied extensively by University of Michigan professor C. K. Prahalad.[10] He points out that, on a global scale, the four billion people at the bottom of the economic pyramid, with purchasing power parity below $1,500 per year, have collective assets and income that exceed the annual GDP of their respective countries. The informal sector produces over 60 percent of the GDP in a typical developing country; the value of the homes in informal (illegal) settlements is as much as 69 percent of the total long-term external debt.[11]

Considering their enormous numbers, the poor control a significant part of the income and assets of a country like Nigeria. Their spending

---

[10] C. K. Prahalad, *The Fortune at the Bottom of the Pyramid: Eradicating Poverty through Profits*, Upper Saddle River, NJ: Wharton School Publishing, 2005.

[11] Hernando de Soto, *The Other Path: The Invisible Revolution in the Third World*, New York: Harper and Row, 1989.

power far exceeds the value of World Bank loans to their countries, and a small part of it could purchase the services, such as safe water, medicine, and electric power, that the government is failing to provide, despite repeated investment by the World Bank and other donors.[12] The problem, however, is that most poor people in developing countries have no formal title to their assets and are thus unable to draw or borrow on them. The poor consume nearly all of their income, and they have no reserve or savings to invest in services and products that will bring them and their children health, education, and future income. Financial services such as reliable savings institutions and secured personal loans are not available to them.

To tap into the wealth at the bottom of the economic pyramid, according to Professor Prahalad, it is necessary to create the capacity to consume. "Cash-poor, and with a low level of income, the [bottom of the pyramid] consumer has to be accessed differently."[13] He lists three principles that govern the capacity to consume:

1. *Affordability*. This principle may involve low-cost, single-serve packages (e.g., of food, cosmetics, kerosene, cell phone calls), novel purchasing or leasing schemes that do not compromise quality or efficiency (e.g., for safe water, recharged batteries, furniture), or microcredit (e.g., for water filters, photovoltaic systems, building supplies).
2. *Access*. Products and services must be distributed near homes or workplaces after working hours.
3. *Availability*. Products must be available and deliverable when people have cash, on paydays or market days.

One more principle particularly relevant to the products discussed in this report is *maintenance*. Complex products must include an on-site service contract to ensure that the product remains functional and that maintenance services and replacements of filters, batteries, and such are performed on time.

There are many examples of new products and services designed especially for the poor that have been successful in some countries. The list includes fingerprint-coded Automated Teller Machines, cell phone access by the call, and staffed computer kiosks or community centers to provide information from the Internet, such as legal documents, market prices, and educational materials. These examples are based on expensive,

---

[12]The World Bank currently has a project portfolio in Nigeria of $1.8 billion, including $320 million in the water sector. See http://web.worldbank.org.

[13]Prahalad, *Fortune at the Bottom of the Pyramid*, p. 16.

capital-intensive facilities, and they can be justified economically only by the sheer numbers of potential customers.

A further potential benefit of marketing to the poor: "When the poor are converted into consumers, they get more than access to products and services. They acquire the dignity of attention and choices from the private sector that were previously reserved for the middle class and the rich."[14] This benefit translates into price and quality competition, which is sometimes particularly intense for items that are purchased in small quantities every day. It also requires firms to build trust in the consumers, a commodity not usually bestowed on the poor.

Professor Prahalad goes on to suggest twelve principles of innovation for bottom-of-pyramid markets. These principles illustrate the kind of approach necessary to succeed in mass markets in developing countries:

1. Focus on the price performance of products and markets. High quality at lower price is required for sustainable success.
2. Innovation sometimes requires hybrid solutions using advanced technologies blended with existing infrastructure.
3. For these large markets, solutions must be scalable and transportable across countries, cultures, and languages.
4. Innovations must reduce, conserve, and recycle resources, including especially packaging.
5. Functionality is more important than form. Products designed for washing clothes in a running stream require a different capability than those for color-separated clothes in a modern washing machine.
6. Process innovation must take into account the realities of the logistics infrastructure for local manufacture.
7. Design of products must take into account the low skill levels and poor infrastructure in poor areas.
8. Education of customers on correct product usage is critical. Often, the usual media outlets such as radio and TV are absent, and new methods must be developed. The vendor has the principal role in this.
9. Products must be designed to work in hostile environments, including irregular power supply and contaminated water.
10. Consumer interfaces, such as advertising, consumer education, and operating instructions, must take into account the heterogeneity of the consumer base in terms of culture, language, and educational and skill levels.
11. Innovation must include methods of distribution and marketing. The extremely dense urban environment and highly dispersed rural communities that characterize the bottom of the pyramid require new, and

[14] Ibid., 20.

possibly advantageous, approaches. The advantages might include a high degree of community organization and a tradition of sharing and reinforcement of ethical behavior.

12. New products and services not required in developed countries should be considered for use in developing countries. These include household water purification systems and home photovoltaic units to provide electricity.

Microfinance

A common problem for business models that involve poor franchisees or customers is the need for credit. Poor people are generally defined as those with low incomes and few fungible assets. Thus, they are unlikely to have savings available to purchase consumer items such as water filters, solar electric systems, or televisions. Poor would-be or actual entrepreneurs may find it impossible to invest in the necessary space and materials to start a business. For over two decades, the concept of microfinance or microcredit has filled this gap, providing capital for simple to complex businesses in the developing world.

Microcredit is the extension of very small loans to unemployed or poor entrepreneurs and others living in poverty who are not *bankable*. These individuals lack collateral, steady employment, and a verifiable credit history, and therefore cannot meet even the most minimum qualifications to gain access to traditional credit. Microcredit is a part of microfinance, which is the provision of financial services to the very poor. Apart from loans, it includes savings, microinsurance, and other financial innovations.

The concept of microfinance originated in Bangladesh in 1976 through a pioneering experiment by Muhammad Yunus, a professor of economics.[15] In 2006 Yunus won the Nobel Peace Prize for his work in microfinance. In the form pioneered by Yunus and his Grameen Bank, microfinance involves the concept of "joint liability." A group of individuals (almost always women) join together to form an association of borrowers. In India, for example, the members of the so-called self-help groups (SHGs) then undergo a training program on the basic procedures and system requirements of borrowing. Loans to individuals within an SHG are approved by the other members of the group, who are also jointly responsible for loan repayment. The members of an SHG save regularly. To minimize the financial burden, limits are placed on the amount lent, and the repayment is typically over 50 weeks. Individuals not in a group

[15]Grameen Bank, 2003, http://www.grameen-info.org/mcredit/.

also can receive microloans. The overall payback rate of microcredit loans is 97 percent.

For the types of businesses discussed in this report—that is, ones that provide basic needs—the consumer purchases are either continual, such as safe water acquired from community water services; intermittent, such as malaria therapy; or capital items with long service lives, such as home solar electric systems. In the continual or intermittent cases, the types of microsavings plans pioneered in Bangladesh and India would provide a cushion to ensure continued access to safe water or antimalarial drugs. For capital items, it would be necessary to go beyond the traditional microcredit schemes to arrange loans secured by the capital item itself and supported by maintenance and service contracts. In some countries, a targeted educational campaign by the vendor has made it possible to persuade commercial banks to provide these loans to poor buyers at commercial rates.

Franchising

Another concept linked to technology transfer, business development, and finance is franchising. In developed countries, franchising is exemplified by retail and service industries such as McDonald's restaurants and Radio Shack retail stores in the United States. Franchising involves an agreement between a *franchisor*—an established national or international company with technological know-how, a known and branded product or service, and a successful business model—and multiple *franchisees*, which are local small companies or entrepreneurs. The franchisees receive training, technical assistance, and a business plan, and they lease or buy technology and materials or products from the larger company. They usually operate under the brand of the franchisor, appearing to be branches or subsidiaries, although in reality they are independent businesses that may be bought and sold by their owners, with the agreement of the franchisor. The more common arrangement between companies based in developed countries and companies in developing countries is usually called a joint venture. In joint ventures, the larger and established companies in the host countries with facilities and their own brands contribute to the partnership. The developing country partner does not usually operate under the brand or name of the developed country partner, although the host country partner may manufacture or sell the latter's branded products, such as automobiles.

Franchising can have multiple benefits in developing countries. Companies that have successful products and business models in developed countries would benefit from the experience, location, native language skills, and local market knowledge of the franchisees. The franchisees benefit from the access to technology or know-how, training, and the brand and reputation of the products. This model has already been put in place by several international companies, including McDonald's and Radio Shack.

An expansion into other products and services besides retail and food would provide new profitable markets for established companies and essential services for the developing countries. In one example described in Chapter 3, a U.S. company set up a subsidiary in the Philippines that, in turn, sold the proprietary technology, with complete supporting services, to 50 branded water stores in the Manila area. The association with a successful brand and the support of the franchisor enabled the local entrepreneurs to obtain loans, secured by the equipment and the business itself, from commercial banks.

In summary, it has been amply demonstrated that marketing products to the four billion poor people on the planet can be profitable for private enterprises, both multinationals and national companies, small and large. Their products range from construction materials and medical services to cosmetics and electronics.[16] The business model is often different from that for selling to the middle class, but it has been mastered by a variety of companies. The experience in offering basic services such as electric power, safe water, and medicines to the poor at a profit is less well known, but the actual examples described in Chapter 3 provide a model that can be adapted to a country like Nigeria. Incentives to companies to profit are, in principle, not required, but, in practice, the risks are not well understood, and the government or donor groups can encourage and expedite the investment by offering partial subsidies and other incentives. Such incentives are discussed in Chapter 4.

A final point is that the types of enterprises discussed in this report are labeled science-based enterprises because they are dependent on exploiting technologies or processes based on the results of scientific research. These technologies are subject to continual revision and improvement in order to remain competitive. The four specific inventions examined here (two for water purification and one each for solar energy and malaria treatment) are all less than 10 years old and are still undergoing improvement. The companies that sell these products have staffs of technically trained people and maintain a certain level of technical competence. The same is true, of course, of the companies, such as electric utilities, municipal water facilities, and pharmaceutical companies, that provide equivalent services to the middle class. The products designed for the poor are not inferior to the common version, but they may be technically different and require dedicated engineering. Descriptors commonly

[16]See Prahalad, *Fortune at the Bottom of the Pyramid*, for a review.

associated with such products are "low cost," "locally manufactured," "durable," "easily maintained." They are safely operable by illiterates and not easily counterfeited with inferior products. They are often referred to as "appropriate technology." These are not trivial requirements, and the engineering design for these products can be extremely challenging. For this reason, they are of specific interest to the academies of sciences of both Nigeria and the United States.

2

Methodology

The methodology used in this study was developed by the U.S. National Academies in collaboration with the World Bank. In the early 1990s, the two organizations worked together on two distinct projects. One was a three-year study in Indonesia, "Science and Technology for Industrial Development," aimed at improving Indonesia's capability to choose and use technologies in its industrial sector. One lesson learned by the two implementing partners was that foreign experts were very effective at identifying opportunities to apply new technologies not generally being used in Indonesia to development problems. However, the tasks of pointing out and avoiding obstacles and designing implementation strategies were much better carried out by members of the indigenous business, scientific, and financial communities.

The second collaboration took the form of organizing a symposium of World Bank staff and National Academies experts entitled "Marshaling Technology for Development."[1] The objective of the gathering was to identify emerging technologies that would be of economic benefit to the client countries of the Bank and would enable them to find new productivity niches in world markets. The lesson for the Bank and the Academies was that the economic potential of a developing country could not be accurately assessed by taking into account only those industries

[1]National Research Council, *Marshaling Technology for Development: Proceedings of a Symposium*, Washington, DC: National Academy Press, 1995, http://newton.nap.edu/catalog/5022.html.

and technologies currently existing in the country. Inevitably, emerging technologies could greatly affect the economic future of a country, either because they offer opportunities for exploitation, such as information technologies, or because they represent challenges and competition that must be met, such as bioengineered substitutes for natural products produced by the country. The Bank sought a methodology for assessing some of these technologies in the context of the economic, social, and cultural environment of the country, which led to the development of the "knowledge assessment." The National Academies produced a version of the methodology in 1996, and it was field tested in Prince Edward Island (PEI), Canada, in 1998.[2]

The knowledge assessment methodology was designed to draw on the lessons that emerged from the Indonesia and "Marshaling Technology" projects. It was intended to be used in small, relatively homogeneous developing countries and to be carried out by an organization providing foreign expertise in collaboration with a national institution, such as a university or nongovernmental organization (NGO). The participants in the PEI knowledge assessment described it this way:

> Knowledge Assessment is based on a venture capital model, and involves some role-playing both by the foreign visitors and the local participants. The visitors act as the agents of hypothetical venture capitalists with a potential interest in investing in the local economy. The local participants act as consultants and possible partners anxious to help the investors find the most promising areas in which to invest. They also want to point out areas where there may be a perceived weakness in the economy or in the knowledge available so that remedies can be taken that would encourage investment. In later phases, local participants play the roles of stakeholders in the enterprises selected by the investors and help them prepare business plans. In each case, of course, the object is not to select enterprises for real investment but to use this method as a vehicle for exploring the strengths and weaknesses of the local knowledge economy.
>
> The methodology itself has its strengths and weaknesses. Its strength is that it is relatively rapid and low cost. It draws on the knowledge of insiders through strategic questioning by expert outsiders. Its weakness is that the information generated is no better than what is offered by participants (and understood and digested by the visitors), and superficiality, bias, and wishful thinking are risks. It is not an exact science, but

---

[2]National Research Council, *Prospectus for National Knowledge Assessment,* Washington, DC: National Academy Press, 1996, http://newton.nap.edu/catalog/9528.html; National Research Council, *Lighting the Way: Knowledge Assessment in Prince Edward Island,* Washington, DC: National Academy Press, 1999, http://newton.nap.edu/catalog/6413.html.

a practical method of eliciting the knowledge possessed and often not explicitly recognized by the local stakeholders.[3]

HYPOTHETICAL CASE STUDIES

An important stage in a knowledge assessment is the development of hypothetical case studies involving cutting-edge technologies. (In the original formulation of the methodology these were called virtual case studies.) A workshop is held for each of the selected technologies. The workshop produces a business plan or implementation plan for creating an enterprise that exploits the particular technology in the local economy, in this case Nigeria. The workshop is led by two or three foreign experts who have experience founding or managing a successful enterprise based on the technology in their own countries, often at the level of CEO. Generally, they have little knowledge of the host country. The other participants are selected from the local business, technical, financial, and educational communities for their knowledge of the technology, the market, labor, or finance.

The format of the workshop is role playing. The foreign experts play the role of investors who wish to create such an enterprise in the country and prepare a business plan. The local participants assist them by playing the role of a board of directors or consultants and providing local information in response to their questions. This question-and-answer mode usually begins by a foreign expert describing what must be done to create a successful business in his or her own country and then asking, "How can we achieve this here?" Very soon the local experts begin asking the questions as they take into account their own different situation. The hypothetical enterprise becomes very specific as the participants decide on a name, site, number and quality of employees, recruitment strategy, market strategy, and choice and source of the technology. And they estimate the costs. Most important, they identify obstacles and propose solutions to problems. Sometimes, these solutions require actions by government, private sector associations, and other institutions, and these actions are reflected in the recommendations of the report. Often the same recommendations emerge from different workshops exploring very different technologies, and these recommendations can be considered high priority for a government or society that wishes to encourage science-based enterprises in general. Overall, the workshops tend to be both highly instructive and provocative for the local participants, and sometimes the hypothetical enterprise takes life from participants in the workshop, without further assistance from the organizers or the government.

[3]National Research Council, *Lighting the Way*, p. 9.

In some instances, there may be a distinction between the opinions of the participants in the hypothetical case studies, which are reproduced in the appendixes to this report, and the views of the authoring committee. Sometimes, the committee has had access to more recent information. In any case, the workshop reports represent the views of those named in the appendixes, and the committee has derived much useful source material from them in preparing the report and recommendations.

As noted, for this project a separate workshop was held to explore each of the selected technologies. The solar energy workshop was held December 8–9, 2005, in Lagos, Nigeria; the safe water workshop was held December 12–13, 2005, in Lagos; and the malaria therapy workshop was held April 24–25, 2006, in Paris, following a preliminary session December 5-6, 2005, in Ibadan, Nigeria. Nearly all of the information on the technologies in this report was provided by the expert participants in the workshops. For this reason, the emphasis in this report is not on providing technical details, which in any case must be checked and updated, but on describing the range of technological choices and the questions that must be asked by potential entrepreneurs and investors.

NIGERIAN ADAPTATION

The motivations underlying application of the knowledge assessment methodology in Nigeria differed from those underlying the methodology presented to the World Bank. No attempt was made to select technologies because of their potential for igniting the economy as a whole, and Nigeria is in fact much larger and more complex and inhomogeneous than the countries contemplated by the World Bank for application of the knowledge assessment methodology. The Nigerian Academy of Science was interested in the problem of providing certain basic services to sectors of the population that lacked them. With some early advice from the Chinese Academy of Sciences, the U.S. and Nigerian academies selected electric power for rural communities, safe water, and malaria treatment for consideration. These are services and products that arguably a majority of the population lacks, and the technological solutions needed to provide them are well known and available off the shelf in some other countries, but they are not available at affordable prices in Nigeria. Even though the technologies were determined in advance, application of the hypothetical case studies segment of the knowledge assessment still appeared to be valid. Furthermore, the three technologies are quite different in the type of businesses or franchises involved, as are the complexity and cost of the technology.

Solar photovoltaic systems are installed in the home or community to provide electricity. At present, the technology must be imported into

Nigeria, but the service can be offered by a small local company. Maintenance, always important, must be provided on-site and regularly. With proper consumer credit, such systems can be made affordable to poor homeowners and small rural businesses.

Safe water can be provided in many ways. The hypothetical case study considered ceramic filters that can be operated in the home and the more elaborate ultraviolet (UV) filters that can be purchased by a community or franchisees who would then sell safe water to households. Active operation of the unit is required, and therefore training is an important part of the service. The price of the water to the consumer would be low and thus affordable to most families without assistance, and the savings on medical services would be substantial.

Although *artemisinin-based malaria therapy* is needed worldwide, it is presently available at low or no cost to patients only in some government clinics in the developing world. That situation will soon change, however, because of the international actions described in the next chapter, and Nigerian companies may have an early opportunity to become part of the global value chain while ensuring that Nigerians have reliable supplies. The Nigerian government has already stated its commitment to this end.

3

The Case Studies

SOLAR PHOTOVOLTAICS

Nigeria receives a large income from oil production and the export of natural gas to neighboring countries in West Africa. In 2001 petroleum consumption accounted for the lion's share of Nigeria's total energy consumption—61.4 percent. Natural gas accounted for 31.7 percent, hydropower for 6.8 percent, and coal for 0.2 percent.[1] However, Nigeria's electric power network serves only 36 percent of the population, mostly in urban areas and often intermittently. This lack of service has a significant impact on nearly all of the country's development goals. In urban areas, which are mostly served by the grid, frequent power breaks affect the viability of existing industries and the development of new ones. Expensive private generators are frequently the only recourse. Small businesses and homes in poor neighborhoods are often bypassed by the grid, and power theft is a common problem. In rural areas, unlit homes and communities prevent children from completing their homework at night and thus have a negative impact on education. The lack of TVs, radios, refrigerators, stoves, cell phones, and other electrical devices reduces the quality of life and opportunities for small business, and results in the uncontrolled migration of potential agricultural producers to the cities. Lack of an energy-dependent infrastructure, including communications,

[1] U.S. Department of Energy, "Country Analysis Briefs, 2003," http://www.eia.doe.gov/emeu/cabs/nigenv.html.

transport networks, and health services, further discourage the growth of local economies.

Solar photovoltaic (PV) systems are renewable energy sources whose application globally has been limited primarily by the intensity and duration of sunshine in the places in need (and the related availability of the land needed to set up collectors for large-scale applications). The cost and efficiency of PV cells are yet another problem. In recent years, the cost of such systems has fallen along with the cost of purified silicon, the semiconductor at the heart of a solar cell, and the efficiency of manufactured solar cells has risen, with the result that the cost of generating solar electricity is only slightly higher than the average cost of power from the grid. Solar photovoltaics may not be ready to generate electricity for the grid, but it can be the technology of choice for communities that lack access to the grid or where the grid is unreliable. A large part of Nigeria is in that category most of the time, and yet the government finds that many Nigerians do not consume enough power to justify the expense of extending the grid into rural or isolated regions. Small-scale solar PV systems may provide the solution, if it can be demonstrated that, with certain incentives, private enterprises can fulfill that need, while ensuring sustainability by making a profit and ensuring efficient use by enabling consumers to afford the cost.

Although developing countries now have a great deal of experience with the installation of solar home systems, there are few places in which an extensive residential area is illuminated for an extended period by solar PV. It is well known that most of these installed systems fail. The principal reasons are not technical, but failures of the business model and poor adaptation to local customs and capabilities. These failures include the lack of training for users, the lack of service and maintenance, and the lack of ownership felt by the users. The last is characteristic of systems that are donated by universities or nongovernment organizations (NGOs) for demonstration purposes. These systems typically are in operation for less than a year.[2]

Solar photovoltaic systems suitable for rural households—solar home systems—usually consist of several components (see Figure A-1 in Appendix A). They include a PV module containing the silicon cells to be mounted on the roof or another sunny spot, a battery for storing electrical energy for use at night, a charge controller, wires and structural frames, and outlets for lights and other appliances. Such a system can operate several fluorescent lamps (often four), a radio or black and white television, and perhaps a fan. The system normally operates on 12 volts, direct current. Long-lasting, deep-cycle batteries, which can discharge

[2]Mark Hankis, "Fresh Ideas Needed: Building the PV Market in Africa," *Renewable Energy World* 9 (September–October 2006):103–115.

80 percent of their charge during extended overcast weather, are best, but automobile batteries, commonly available in Nigeria, also could be used. The charge controller prevents damage to the system in the event of overcharging by the solar module or prolonged battery discharge from overuse. Other requirements are installation, periodic battery replacement (once every five years for a deep-cycle battery), and user training; they are often part of a service contract for maintenance. The cost of a 40-peak watt system is about $350–$500 worldwide, *depending largely on the input duties on the solar panel*, but this cost is beyond the reach of most Nigerians whose annual per capita income is about $250. Further complicating the situation, on August 31, 2006, the street price of kerosene, widely used for cooking and lighting, was raised to 650 naira, or about $4.60, for 4 liters (gallon). With this price increase, kerosene, which is mostly imported, became the most expensive petroleum fuel in the country. As recently as 2003, the price had been $0.78 for 4 liters. The alternative for many families is firewood, and so Nigeria's forests were put at greater risk by the government's increase in the price of kerosene.[3] That price will be a key indicator of the willingness of people to pay for solar electric systems for their homes.

The task of the hypothetical case study workshop was to determine whether such a system could be made affordable for Nigerians. To do so, the workshop studied the Solar Electric Light Company (SELCO) in the state of Karnataka, India (http://www.selco-india.com). SELCO was founded in 1995 with initial financing from the Rockefeller Foundation. It was the first rural solar company in India to be engaged wholly in designing, marketing, and servicing a wide range of solar-powered equipment and installations for lighting, TV and radio, water pumping and purification, and many other applications.

The members of the workshop panel represented business, government, and academia. Two of the participants are directors of successful Nigerian companies that sell solar photovoltaic units, one established by Siemens Corporation and the other self-started. Both rely on large contracts from hotels and other large entities and appear to be highly profitable, but neither is at present serving rural villages. Two university groups that have developed solar PV units and donated them to villages also were represented. The head of the National Agency for Science and Engineering Infrastructure was present as well. He described his ongoing efforts to develop a critical mass of Nigerian solar specialists and to establish by 2007 a solar photovoltaics cell manufacturing facility. Funds for the facility have been released by the government, and Princeton University is assisting with the training in silicon amorphous film deposition.

---

[3]The official price is now 53 naira per liter, or $1.50 per gallon, but kerosene at that price is very scarce. The street price is three times higher.

BUSINESS PLAN FOR SELCO IN INDIA

A key element of the SELCO business plan is linkage with rural banks that will provide loans for PV systems. Under their Solar Lighting scheme, these banks are now offering consumers three- to five-year loans for 90 percent of the solar unit cost at an interest rate of 12–12.5 percent, which is below commercial rates. SELCO assumes complete responsibility for performing all the other tasks: (1) organizing awareness campaigns in rural areas, including demonstrating PV systems; (2) identifying and prequalifying potential beneficiaries; (3) training local technicians, installers, and service personnel; (4) installing solar home systems purchased through the lending bank; (5) educating users; and (6) providing after-sale service and maintenance.

SELCO offers consumers a "lease to own" scheme in which the consumer pays a quarter of the total system cost, including service and installation, as an upfront payment and receives the rest as a loan. SELCO procures systems from reputable manufacturers only after securing factory guarantees of quality, which are passed on to the consumer as performance guarantees. It has set up branches in villages, and its teams of local technicians on motorcycles ensure quick after-sale service and regular collection of loan installments. A typical SELCO branch is staffed by its own technicians, salesmen, and collection agents, all hired locally. The technicians work for SELCO on a salary and commission basis, and thus they have an incentive to sell more systems. The company now has 170 employees in 25 centers and over $3 million in revenues a year.

Reaching this point did not come easily. Initially, solar units were installed at no charge in prominent locales such as the house of the local village chief and the local religious building. These systems acted as demonstrations to other villagers and local financial institutions. From its earliest years, SELCO realized the importance of consumer financing, and it spent much of its human and financial resources on informing bankers about the usefulness of solar technology. In four to five years, SELCO was able to convince more than 550 managers of seven different local banks of the value of financing solar home lighting systems. After SELCO conducted several rural bank sensitizing programs—such as training bankers in technology assessment, arranging demonstrations on bank premises, and holding bank and customer meetings—bankers' confidence about financing solar home systems steadily increased. In 2002 the cost of household systems was $600, the same as a motor bike. Now, after efforts to open new markets and the appearance of competition from other companies, the cost is about $400. At present, SELCO has 50,000 customers in the state of Karnataka and agreements with all the banks in the state to finance solar electric systems, based on no collateral except the system itself. More details on SELCO and the hypothetical Nigerian company described in the next section appear in Appendix A.

SOLAR HOME SYSTEMS IN NIGERIA

After looking at SELCO, the workshop considered a hypothetical enterprise called the Solar Energy Company of Nigeria Ltd. (hereafter the Company). After reviewing all the elements, the participants concluded that in Nigeria a 40-watt system could be produced and sold for about 75,000–80,000 naira, or about $500–$600. It would be designed to provide about four hours of light at night when there is adequate sunshine in the daytime. For an additional $250, a 50-watt system with an inverter could also support a refrigerator. At present, no companies in Nigeria are providing a similar product, although several university-based NGOs have installed home solar systems for free in villages on an experimental basis. The only major competition is the Power Holding Company of Nigeria (PHCN), which maintains the grid but does not provide power to the villages. However, under law it is the sole supplier of power to homes and businesses, and so it could challenge the legality of a solar energy company.

The Company, like SELCO, would have to import, at least initially, the solar module containing the silicon wafers, but, as the local industry develops, it will be more economical to buy them locally. The same is true of the deep-cycle battery. The factory offers a three-year warranty on the battery, but the Company may have to offer the banks a five-year warranty, which will require careful maintenance of the batteries. The inverter is already available locally.

Nigeria has experience and core competency in solar energy. Many small companies are serving niche markets, including solar water heating and drying, and several universities have ongoing research programs. The potential market is huge, possibly up to 100 million people who are underserved by the national grid. A potential local problem in some areas is that users may become discouraged during the rainy season when sunlight is in short supply for several months. A program of battery exchange may be a useful complementary service during such periods.

The Company could begin operations with as few as five persons, plus two-person teams of installers with cell phones and motor bikes able to complete two installations plus sales per day. Other requirements are a central office space for storing materials and a small showroom. Marketing could begin simultaneously in an urban area, where paying customers and credit are more plentiful, and in a selected rural area with good road access. In rural areas at least, buyers will require microfinancing. The Company will have to plan a campaign with banks on behalf of buyers that includes demonstration of the system and the existing guarantees. The SELCO experience in India, described in Appendix A, would be instructive. The Nigerian Association of Small and Medium Enterprises (NASME) has a microcredit arm that could be useful. Usually, this microcredit is directed at start-up businesses, but the argument could be made that electric lighting would give consumers an opportunity to undertake work or conduct a small business in the home to repay the loan. In addition, the government could offer consumers incentives such as tax deductions or free service contracts for investing in renewable energy sources that would subsidize part of the purchase.

FINANCING THE ENTERPRISE

Although any individual, entrepreneur, investor, or company already engaged in other aspects of solar energy or of service delivery in rural areas could establish a solar energy sales and service business, the major problem facing most of them will be first-stage financing. Fortunately, several sources in Nigeria might have a particular interest in businesses of this sort. For example, the Small and Medium Industries Equity Investment Scheme, or SMIEIS, offers equity loans under the condition that SMIEIS takes an equity position in the borrowing firm, which it can convert or sell after a fixed time period. SMIEIS was created by the Committee of Bankers in Nigeria on the premise that all Nigerian banks would agree to contribute 10 percent of their before-tax profits to provide something like venture capital for small and medium-scale Nigerian industries. SMIEIS is intended to stimulate economic growth and development, develop local technology, and generate employment. The participating banks for equity investment under this scheme have currently set aside over 5 billion naira for an alternative approach to financing small and medium-size enterprises (SMEs).

The government of Nigeria must, however, provide some form of incentives that would encourage the private sector to make a significant contribution to providing electric power to rural areas. Such incentives could include some form of loans or loan guarantees for start-ups, lower tariffs on goods imported for solar systems, price subsidies for consumers, or coupons that help enable consumers to pay for service contracts on installed units. (If the total cost is subsidized, the service contract should be included in the base price. If the service contract alone is covered, it could be a separate item. In either case, the service contract should not be optional. Certainly, the banks will require it.) The incentives should not include direct payments or contracts to companies to provide solar electric systems at no cost to consumers. Experience worldwide has demonstrated that such an approach is a formula for corruption, noncompliance, nonsustainability, and a short lifetime for the installed system.

WATER PURIFICATION

Life expectancy in Nigeria is 46 years. Water-borne diseases account for the second largest loss of disability-adjusted life years (DALYs) after malnutrition, and a large part of the disease burden is diarrheal disease, most seriously in children. Only 60 percent of households have access to improved sources of drinking water,[4] although it is widely believed that no community in Nigeria receives safe drinking water from the government. This situation affects both rich and poor; even those who rely on deep bore wells cannot reliably secure uncontaminated water.

Of Nigeria's 320 million cubic meters of water, 86 percent is surface water. Forty percent of the population, mostly in rural areas, has access only to surface water. The water table ranges from 300 meters in sedimentary areas to 70 meters in basement areas. Water can be found above 70 meters, but with much saline intrusion.

In Nigeria, water is sold on the street in containers ranging from sachets in plastic sacks to bottles of mineral water, and the quality of the water varies widely despite the efforts of the National Agency for Food, Drug Administration and Control (NAFDAC) to control and license all providers of water to the public in plastic sachets or bottles. Water sellers often get their water from wells. Some is boiled or treated with ultraviolet radiation. Some facilities are inspected, but the water is not tested. Street water sells for 5 naira ($0.04) for a 500-milliliter bag and 40–60 naira for a bottle. Commercial bottlers can get water from industrial steam as well as from natural springs. But even bottled water is sometimes contaminated by users after purchase by, for example, adding locally made ice. In Nigeria, demonstrably safe water is more expensive than petrol. The health costs of not having safe water are even more expensive.

For many rural populations, there has never been an alternative to drinking contaminated and sometimes turbid water. As a result, many people have never had clean water. Some potential users argue that the resistance developed by people to water-borne diseases will be lost if clean water is provided. But children under age five have no resistance, and many people suffer from diarrhea. Education will play a critical role in the adoption of safe water technology and handling practices in the communities served. When people understand the high costs associated with drinking unhealthy water, including ill health, death of children, loss of working hours, and hospital fees, they will realize the economic

[4]UNICEF, http://www.unicef.org/infobycountry/nigeria_28236.html.

benefit of paying a small price for treated water. This realization will provide an opening for private companies to enter the market for safe domestic water.

Technologies for Water Purification

Many effective technologies are available for home-scale water purification. Some are chemical-based, such as treatment with chlorine; some are thermal, such as boiling and solar heating; some are mechanical, such as ceramic filtration; and some utilize radiation, such as ultraviolet (UV) treatment. Many of these technologies are essentially free, such as the SODIS system developed in Switzerland, which requires only a blackened used soft drink bottle and four hours on a sunny roof, although they may require extensive training for users to acquire the discipline necessary for health protection. Indeed, discipline and sustainability are key elements of success in providing safe water to rural households. Sustainability implies freedom from the exigencies of the government public works process and the vicissitudes of donor financing. The market for an essential commodity such as safe water will never become smaller; to the contrary, it will increase as the population grows, and in it lies an opportunity for the private sector to sustainably fill this need.

The hypothetical case study examined two different technologies that have been successfully exploited by private, profit-making companies in other parts of the world: mechanical filtration using specially prepared ceramic vessels and mechanical filtration combined with ultraviolet irradiation. Both technologies have their own competitors—other mechanical filters such as cloth (the least expensive and most affordable technology for remote villages) and different designs of UV filters—which lends some flexibility to the application in Nigeria. The ceramic filter, called Filtron, was developed in Nicaragua by Potters for Peace (http://www.pottersforpeace.org). It is intended to purify drinking water for home use. The cost is very low, but consumer training and discipline are important for proper application. The UV filter, developed by WaterHealth International (http://www.waterhealth.com) in California for developing countries, is applied on a community scale to produce potable water for all domestic uses except laundry. It is marketed to small commercial water sellers and village cooperatives in Mexico and India and to franchised "water stores" carrying the WaterHealth brand in the Philippines, all of which have been able to pay off their loans and make a profit.

For each technology, the important element is the business plan for selling safe water to poor consumers. Other technologies could also be used with similar business plans. One example is the "Bring Your Own Water" system installed in Muramba, Rwanda, by Engineers Without

Borders–USA.[5] Other technologies employ parabolic mirrors to increase the radiation dose or use SODIS after filtration. Simple, low-cost technology such as using sari cloth for water filtration has shown to be effective in removing cholera bacteria from water in Bangladesh. Other, more advanced technologies based on nanotechnology are still under development and showing promising results.[6] The use of each technology, however, must be accompanied by education about the value of safe water and proper maintenance.

Ceramic Filters

The Filtron ceramic filter is intended primarily for household use, ideally as part of an overall water delivery system. The most economical filter consists of a porous clay pot perched inside a lidded 5-gallon spigoted receptacle made of plastic or clay. The pot is saturated with colloidal silver, which acts as a germicide/disinfectant. The simplest unit has a flow rate of about 2 liters of water per hour, which is enough to provide a family of five to six persons with drinking water. Another model processes 6 liters per hour. The filter has been laboratory tested successfully in over 10 countries on four continents, and it has been pronounced effective in eliminating coliform bacteria, parasites, amoebae, and *Vibrio cholerae* from water. Some 100,000 Filtrons are in use throughout the world, serving about 500,000 people.

Potters for Peace exclusively trains ceramicists in developing countries to make Filtron filters for a small fee; there is no license cost. The cost of the product depends mainly on the local cost of labor and electricity or fuel, and it varies between $5 and $25. In Nicaragua, the cost is $7, and the Filtron lasts about two years. A designer model of the Filtron is available for up to $100. It features the same water-purifying effectiveness in elegant containers designed for a more affluent clientele, which opens up the possibility of manufacturing for a second market of consumers.

The cost of a factory that employs two to four persons is about $10,000, which can be recovered in as little as nine months. Potters for Peace offers the plans of the press and other equipment free of charge, or it will refer the client to mechanics who build a press to order at a cost of $800 for a portable system able to produce about 50 clay filtering elements a day. Usually the producer is already in the ceramics business and so may already have on hand the equipment and skilled personnel.

---

[5]Engineers Without Borders, http://www.edc-cu.org/pdf/EWBRwandaJuneSummary.pdf.

[6]Thembela Hillie et al., "Nanotechnology, Water, and Development," Meridian Institute, Washington, DC, 2003, http://www.merid.org/nano/.

At the conclusion of the hypothetical case study, Ron Rivera from Potters for Peace visited the ceramics laboratory of the Federal Institute for Industrial Research in Lagos, and in one day he and lab personnel made two prototype Filtrons. A diagram of the Filtron is shown in Figure B-2 in Appendix B.

Ultraviolet Filtration

The patented UV Waterworks (UVW) unit is at the core of systems sold by WaterHealth International (WHI) to franchisees that produce and sell potable bottled water to consumers at prices below those of packaged bottled water. Other WHI products, such as community water systems, are sold to governments or communities directly, and provide enough safe water to meet all daily domestic needs, including hand and food washing and bathing. Operating costs for a system that can serve at least 3,000 people are less than $4 per person per year. Treated water is sold to recover the investment and maintenance costs of these systems at prices that are within reach of the populations being served (see Figure B-1 in Appendix B).

One of the features of the UVW system is that the UV source is suspended above the water being treated rather than being submerged in the water, where it would be exposed to corrosion and biofilm formation. Water passing through the system is irradiated at high intensity amplified by reflection. Its fewer maintenance requirements enable the UVW system to be operated in areas where labor pools may lack technical knowledge or specialized education. Very little maintenance is required—the lamps must be changed annually, and the filters require periodic backwashing and replacement to avoid biofilm accumulation. UVW technology is also designed to be fail-safe. If any type of malfunction occurs, such as a power outage or a drop in radiation dosage, an automatic valve closes the entry port to the device, ensuring that contaminated water cannot flow through the system without being disinfected. WHI's systems can be powered by solar or wind energy or a generator; some units have even been run on a car battery for two weeks in an emergency.

WaterHealth's systems are distributed using several different business models in different countries, all by means of local sales agents. One of WHI's earliest commercial successes was in Manila. In 1997 WHI established a subsidiary in Manila to manage franchises that use the UVW technology to provide lower-cost alternatives to bottled water. WaterHealth Philippines works with entrepreneurs who wish to open water store franchises in Manila, and then sells them the UVW units imported from California. The subsidiary enables local "mom and pop" store owners to own and operate their own WaterHealth-branded water stores. The owners benefit from WaterHealth's expertise (e.g., in terms of where to open their stores and how much foot traffic versus deliveries they should expect) as well as technical services. All franchisees are trained to operate their stores in compliance with the highest sanitary and quality standards. Water is sold from storefronts in sanitary containers. In the thriving Manila market, over 3,000 water stores are now vying for business. The franchisee pays about $8,000 for a turnkey operation plus franchise fees.

An alternative model, the community water system (CWS), can be established as a decentralized "microutility" in areas previously thought to be unreachable. A typical CWS is designed to provide a community of up to 3,000 people with up to 20 liters of safe drinking water per person per day. Systems are modular and scalable—they can be configured easily to serve larger or smaller populations. In India and Mexico, WaterHealth markets the CWS as a "microutility" to governments, entrepreneurs, or village organizations, which recover the investment in the community water systems through the sale of treated water to villagers.

For the purchaser of a CWS, it is a turnkey operation, including water storage and, if necessary, the pipes to bring the water to the village for treatment from a distance of up to 2 kilometers. Profits are made all along the value chain. WHI sells equipment to a local affiliate, who markets the units, with a service contract, to local franchises, entrepreneurs, or village organizations. These groups typically sell coupons to families to redeem for water, and some end users opt to pay extra for home delivery in special containers. Each CWS includes an educational program, usually conducted by an NGO under contract, that encompasses health and hygiene issues and encourages people to use clean water. A standard CWS, with an installed capacity of 65,000 liters per day, can produce enough purified water to meet all the daily needs of a community, provided the water source is available. The cost of this turnkey operation is about $50,000, or about $17 per person, which is lower than alternatives of similar capacity, such as bore wells, that normally provide no disinfection, filtration, safe storage, or education on health and hygiene. By comparison, municipal facilities cost about $100–$250 per person to build, and bore wells are similar. WHI projects that village water boards or local organizations should be able to generate a surplus of $24,000 a year acting as a small utility after communities fully adopt a CWS.

And in Nigeria . . .

Because the workshop considered two very dissimilar technologies, it had to decide whether to consider two hypothetical companies or one. Filtron production is relatively quick to organize at low cost; a WaterHealth–

type operation would take longer and be costlier to assemble. The workshop decided to discuss them together as one company, recognizing that investors can pick and choose the elements that suit real opportunities.

The potential customers of such a company would be any people who lack access to potable water, specifically including those in rural areas for whom the government water service is not available and the urban and periurban poor who cannot afford well or bottled water. Potable water or Filtron units can also be offered to middle-class customers in more elegant containers, and bottled water can be marketed through water stores or by home delivery. Branded water also may be sold in bottles in city traffic, with some security measures applied to distinguish the product from counterfeit (nonpurified) brands or water in contaminated containers.

In rural areas, operators of community water systems or water refilling stations may set up facilities alongside a river or other traditional source, where families would have a choice of collecting water themselves or buying safe water in portable, reusable containers. If the water sales are supplemented with other products and services, ranging from laundry soap to Internet access, the site could turn into a small commercial hub.

The technologies and materials needed to manufacture the Filtron filter are generally found in ceramics facilities everywhere. The filter can be made from locally available clays, wood chips from local saw mills, and rice or peanut husks from local villages. The containers for the water also can be made locally using local facilities for pottery and plastic manufacturing.

The process for making the filter calls for combining the clay with sawdust, peanut shells, millet straw, or a similar material in equal volumes, resulting in 25 kilograms of denser clay to 6 kilograms of lighter sawdust or straw. The mix is then pressed into a mold, dried, and fired in the kiln to 890°C. During the firing process, the sawdust burns off, leaving small twisted pores through the ceramic material too small to permit bacteria to pass. After initial tests of filtration rate, the inside and outside of the filter are coated with colloidal silver, which provides additional antibacterial properties and prevents regrowth. The silver can be imported at a cost of $60 per liter, enough for 500 filters. It might be possible to manufacture colloidal silver in Nigeria, but the production of Filtrons will hardly support a local industry alone.

Of the two technologies considered, the Filtron has the lower cost, but it does require regular maintenance by the household. A biofilm will develop on the inside of the container where water is stored, and so the container must be cleaned periodically. It may be possible to glaze the inside of the container so it could be flushed with boiling water without a significant increase in cost. Experience has shown that poor people will pay for safe water, but they expect a demonstration that it is clean. A low-cost water quality test kit should be included with each filter.

UVW-type units would initially be manufactured elsewhere and imported. It is possible that, after the development of a sufficient order pipeline, the units could be manufactured in Nigeria. The tank, containers, flow meter, pumps, PVC plastic casing, cone and filter paper, cartridge filters, and spare parts could be made or sourced locally. Test equipment or laboratory services also can be arranged locally.

At least 1.5 kilowatts of power are needed to operate a community water system. Depending on the water source, lower power requirements may apply to water store franchises that do not have to pipe water to their facilities. Solar cells and batteries for this purpose can be imported if desired, and in the future they may be available locally.

Maintenance is perhaps more important than the choice of technology. A community water system can be considered turnkey only to the extent that it includes maintenance and service. Once a month the unit must be opened and cleaned with a wet cloth, water samples sent to the lab, and a new certificate of water quality issued, which should be prominently displayed. The UV bulb must be changed once a year. The service organization, which must be based locally, must stock essential spare parts, including UV disinfection units and other parts not available in the country. Production, labor, and legal requirements, including an environmental impact statement for the UV units, are described in Appendix B.

Marketing would likely be done on two different levels. The first would be educational. It will be necessary to make a compelling case, especially to poor people, for the consumption exclusively of purified or filtered water in order to upgrade their quality of life and avoid the negative consequences of poor health. But this approach may be resisted by people who have never had potable water and who do not want to pay, either because they do not see the need or because they believe water is a natural right. It will help to ensure that the vendors are local people and that the emphasis is on purification rather than water per se. A simple analysis of the low cost of treated water versus the avoidably high costs of consuming untreated water should be carried out and conveyed through social marketing and educational campaigns.

The government and NGOs should be enlisted to participate in this educational campaign. An endorsement from NAFDAC would be helpful, if it can be persuaded to admit that the Nigerian population is not adequately served at present. The government should be urged to subsidize diarrhea prevention just as it subsidizes HIV prevention. "Wash your hands, filter your water!" would be a good slogan for a government campaign. NGOs could be involved through the Internet and at health centers, especially maternity and pediatric wards, where NGO employees might be working. Radio can be very effective, reaching a larger population.

The second level of the marketing effort would be branding and identifying the product itself. Scientifically based testing must be emphasized to allow the buyer to distinguish safe water from unapproved and inferior products. (In this context, it is important that the selected technology filter out turbidity in the water to distinguish the appearance from that of untreated, boiled, or chlorinated water without filtration. The effectiveness of the filtering process would then be obvious.)

As for the competition, the association of the companies that sell water in plastic bags in Nigeria should be engaged and employed to distribute safe water or sell Filtrons instead of competing with community water systems. They could also sell low-cost water quality kits, such as the Hach test described in Appendix B.

Nigeria should serve as an excellent market for safe water, because 100 million people are without. Furthermore, local ownership of industries is more widespread in Nigeria than in other African countries. The culture of business and entrepreneurism in Nigeria could lead many to enter this business, including many sellers who presently lack the technology to purify the water they sell. In the case of Filtron, all material and labor, except the colloidal silver, can be obtained locally.

The greatest challenges, however, stem from the very same factors that produce the advantages. The large potential market includes many consumers who have a history of using contaminated and turbid water and who do not understand it is a cause of their health and infant mortality problems. The educational campaign is critical to activate the market. Furthermore, the large number of potential entrepreneurs practically ensures that a short time after a business appears to be successful, or even before, a large number of counterfeit products will appear on the market. For UV-treated water, the difficulty will be significant, because the product will come into direct conflict with traditional water sellers, most of whom do not take care to purify properly the water they now sell, and they will most likely sell at a lower price. Some may mimic the brand. Steps must be taken to clearly brand the product in a way that is hard to imitate, and to emphasize safety and reliability in marketing activities.

Start-up funds can be sought from NGOs such as Water International or Rotary International. Rotary tends to be business oriented, and so it could be helpful in creating new enterprises and preparing business plans. But to introduce filtration and purification on a scale that would serve the most people who are presently outside the water districts, banks or the government must play a clear and significant role in financing new enterprises. Family health extension agents could assist families and communities in providing education, especially information that explains why they need safe water and evaluates the various possibilities for purification.

ANTIMALARIAL ARTEMISININ COMBINATION THERAPY (ACT)

At this time, no developing country government is able to ensure a supply of ACTs for all of its population, because only one company supplies the medication recommended for malaria and there are only a few monopolistic producers of the raw material, and the price is high. That situation, however, is fluid and already beginning to change. Other countries, including several in Africa, plant and harvest *Artemisia annua*. And other pharmaceutical companies, including generic manufacturers in developing countries, should be able to produce ACTs within a year or so, because at least two new coformulations will be available for licensing. At the same time, there is an international effort to subsidize the price at the national level (for more from the hypothetical case study, see Appendix C).

Present Need for Malaria Chemotherapy

Malaria has been a major cause of death in Africa for millennia. During the twentieth century, malaria took 150–300 million lives, surpassing war, famine, and all other infectious diseases. Today, malaria kills 1.2 million people worldwide each year, 1.1 million of them in Africa and 1.0 million of them children under five years.

In Nigeria, 85 percent of the population is at risk of contracting malaria, and 60 million people experience more than one malaria attack per year.[7] It is not uncommon for small children in endemic areas to have four or more episodes per year requiring treatment. Malaria affects the ability of children to learn in school, and even to attend school, and stunts economic growth because of its effect on worker productivity and the time devoted to caregiving. The high malaria incidence in Nigeria puts at risk the potential for reaching the Millennium Development Goals.[8]

Since the seventeenth-century discovery of quinine in Peru, malaria has been treatable, but historically the supply of effective drugs has met the needs of only a small percentage of the sufferers. Introduced in 1945, the synthetic chloroquine became available at low prices in tropical countries in the 1960s and was effective for decades. Eventually, however, parasite resistance to chloroquine evolved in Asia, and it has now spread

---

[7]Communication Initiative, http://www.comminit.com/trends/issuestrends/sld-2098.html.

[8]UN Millennium Development Goals, http://www.un.org/milleniumgoals/.

throughout Asia and Africa. Chloroquine can be purchased for about $1 per course of treatment in much of Africa, but it is increasingly impotent against falciparum malaria, the severest form of the disease. Some other antimalarials have been available for many years, but none of them is as safe and effective—or as affordable—as chloroquine was, and resistance to most of them develops more easily than did resistance to chloroquine. Much work has gone into developing vaccines against malaria. Currently, a small number of experimental vaccines are showing promise, but none is ready for widespread use.

The drug class the world is depending on to carry malaria control into the future is the artemisinins, a family of compounds made from extracts of the plant *Artemisia annua*. *A. annua* had been used for centuries in China against fever, including the fever produced by malaria. At present, where it has been used, artemisinin therapy is safe and effective. It works very quickly and kills several of the stages of malaria parasites—in fact, more than any other antimalarial known, including chloroquine, when it was effective.

In 2001 the World Health Organization (WHO) recommended that oral artemisinin derivatives be adopted as the first-line treatment for uncomplicated malaria. They are also as good as quinine for severe malaria, and, when used appropriately, some are better than quinine for noncerebral, severe malaria.[9] WHO and the global malaria community also recognize the value of using artemisinins coformulated (i.e., two drugs in one pill) with another effective but unrelated antimalarial, in what is referred to as artemisinin combination therapy, or ACT. There are two reasons for using the combination. First, artemisinin derivatives given alone require a seven-day course treatment, but, in combination, the course of treatment can be reduced to three days, which should result in a much higher proportion of people completing treatment. The second and main reason for combining drugs is to inhibit the development of resistance to either drug, similar to what is being done today for HIV and tuberculosis. If the mutant parasite is resistant to artemisinin, it will be eliminated by the other drug. The artemisinin compounds have shorter half-lives in the body and work more quickly than other drugs, thereby reducing the likelihood of a resistant population. Different modes of antimicrobial action will reduce the risk that the parasite will develop resistance to both drugs at the same time. ACTs have been effective in every site they have been tried, including highly endemic areas in Africa. However, ACT is not useful as a prophylaxis because of its short active life in the body. Widespread use of artemisinin monotherapy—that is, using artemisinin alone—could

---

[9]World Health Organization, "Facts on ACTs," http://www.rbm.who.int/cmc_upload/0/000/015/364/RBMInfosheet_9.htm.

encourage resistance and is the biggest threat to the long-term viability of this family of compounds.

The main drawback to ACTs is their price: they are 10–20 times more expensive than chloroquine at their wholesale price, which is $2.40 per adult course of treatment. At the retail level, the price may be 10 times higher. Today, the supply of ACTs is far smaller than the number of sufferers, in large part because effective demand has been limited by cost (even with some financing aid currently available for antimalarials).

In the urban environment, children, who suffer an average of three or four attacks a year, are usually treated with chloroquine. But because chloroquine is ineffective, each incidence may be treated multiple times. When the patient feels better, the treatment is stopped, often prematurely, further promoting resistance. Because chloroquine has a long shelf life, it is saved for the next occurrence. In rural areas, where the incidence is higher, a family could spend up to 40 percent of its income on malaria treatment. Although ACTs are far more expensive, they do not require multiple treatments for the same infection, and they reduce absenteeism from work and school, dedication of time from a caregiver, and the risk of death or permanent neurological damage.

Treatment with ACTs as a public health measure is cost-effective, because they cure a potentially fatal, and more often debilitating, disease for just a few dollars. At present, however, no funds are available from the Nigerian government for the treatment of malaria. Under a new Nigerian government policy, children under five will receive ACTs without cost, but only at public sector facilities. As in other African countries, the majority of people purchase malaria drugs through pharmacies, drug sellers, and neighborhood kiosks. Many, particularly in rural areas, have little access to public facilities.

Part of the reason for the high cost of artemisinin drugs is that much of the supply of high-quality artemisia leaves needed to manufacture the drug is controlled by dealers in China and Vietnam. Chinese companies extract the artemisinin and produce monotherapies and ACTs that are sold internationally without prequalification by international standards, despite WHO recommendations, or they sell the artemisinin directly to overseas buyers.

WHO and several international donors are setting up an infrastructure to distribute ACTs at a subsidized price, but WHO requires the formulation to be manufactured according to a WHO standard based on current Good Manufacturing Practices (cGMPs). To date, only one firm has met the standard, Novartis of Switzerland. It buys the raw material from China and Vietnam and manufactures an ACT with the brand name Coartem. Coartem costs more than $20 for a course of treatment at pharmacies in Lagos. Artemisinins are also sold in blister packs that contain two kinds of pills—artemisinin and another antimalarial drug—which means that the patient can choose to take only the artemisinin, essentially a monotherapy with the consequent risk of generating resistance. The Nigerian government has signaled its intent to ban chloroquine and all artemisinin monotherapies, which may drive the pharmaceutical companies to produce ACTs. However, if the ACTs are not formulated under WHO-approved practices, they may not be accepted in international trade and likely will not qualify for a global subsidy should one come into being.

It is estimated that Nigeria has more than 100 million cases of malaria per year. However, the market for ACTs is generally calculated on the basis of the "real demand," which takes into account national policies and funding available for purchasing by consumers, government, or donor agencies. That figure is about 10 million courses per year for which payment is available, or about 1 out of 10 actual cases. When the supply of ACTs is compared with "real demand" rather than the number of cases, the shortages vanish, and with them the incentive to increase the harvest of artemisia. It is clear that ACTs will never be available to the majority of infected people in Africa until there is global system of affordable supply.

Global Aid for Antimalarial Drug Purchases

The Global Fund for AIDS, Tuberculosis and Malaria provides funds to governments for approved plans covering all types of interventions for these diseases. Assuming funding is available, it will spend $2.2 billion over five years. Global Fund rounds two and four committed to Nigeria about $41 million for malaria control, and up to a possible $130 million over five years, of which $16 million will be used to purchase ACTs. The World Bank plans to disburse $180 million to Nigeria for malaria control in 2007. The (U.S.) President's Malaria Initiative (PMI) will provide 15 countries with $1.2 billion over five years for malaria control, some of which will go to purchase insecticide-treated bed nets and DDT spraying.

In addition to helping governments pay for antimalarials, the World Bank is leading an effort by the Roll Back Malaria partnership to develop the architecture and identify funding for a global subsidy of ACTs, which would be carried out at a very high level (i.e., national frontier or factory gate) and would allow drugs to flow through both the public and private sector supply chains and reach all consumers at low prices.[10] In

[10]Kenneth J. Arrow, Claire B. Panosian, and Hellen Gelband, *Saving Lives, Buying Time: Economics of Malaria Drugs in an Age of Resistance*, Washington, DC: National Academies Press, 2004.

this way, an effective treatment would be available and affordable—that is, it would cost about the same as chloroquine and should cost less than an artemisinin or other monotherapy. ACTs will be sold to governments, NGOs, and private distributors at the heavily subsidized price. Only good-quality, prequalified drugs will enter the market. But, unless the public and professionals receive the information needed to facilitate the switch to ACTs, it is unlikely that the market will grow to include the people with real needs. An important meeting was held on this topic in Amsterdam on January 18–19, 2007. The state of agreement was described in the official summary:

> There was a broad consensus reached during the meeting that the subsidy should be implemented but there were still some legitimate concerns about the subsidy that needed to be further elaborated. However, there was general agreement that careful planning and consultation with partners would ensure that the potential negative implications are minimized and that these concerns were not sufficient to prevent the implementation of the subsidy from moving forward. It was agreed to move forward with the design and that all key stakeholders would remain engaged in this process in the coming months and that the subsidy could be ready for launch as early as fall of 2007. The overall goal of the ACT subsidy will be to increase universal access to and use of ACTs by bringing the price down to the same level as chloroquine and sulfadoxinepyrimethamine and as such diminishing the use of ineffective antimalarials and artemisinin monotherapy.[11]

Where governments turn to international finance to purchase drugs—whether from the Global Fund, the World Bank, bilateral donors, or others—some restrictions are generally imposed to ensure that the drugs purchased are of good quality. All these institutions rely heavily on WHO's standards for cGMPs.

Several Nigerian companies are cGMP-certified for other drugs. "Prequalification" is a newer and higher level of approval developed for AIDS, tuberculosis, and malaria drugs. Prequalification requires meeting cGMPs for the process, as well as demonstrated product effectiveness against the disease.[12] Any Nigerian company intending to produce drugs

---

[11]"Meeting Report," Expert Workshop and Consultative Forum on a High-Level Buyer Subsidy for Artemisinin-Based Combination Therapies (ACTs), Royal Tropical Institute (KIT), Amsterdam, January 18–19, 2007, unpublished.

[12]According to WHO (http://mednet3.who.int/prequal/), the elements of the prequalification procedure are as follows: (1) the manufacturer must provide a comprehensive set of data about the quality, safety, and efficacy of its product, including details about the purity of all ingredients used in manufacture, data about finished products such as information about stability, and the results of in vivo bioequivalence tests (clinical trials conducted in healthy

that might qualify for foreign funds will have to meet these standards, even if the drugs are used only within the country.

The Nigerian government has announced its support for the various WHO initiatives, and it has mentioned a timetable for dropping chloroquine and artemisinin monotherapy from Nigerian markets. President Olusegun Obasanjo of Nigeria has said that he wanted the ACTs market to be driven by the local private sector, and that the government might advance about 40 percent of start-up costs, as was done earlier to encourage the production of cassava by providing cuttings for planting at a subsidized rate. The government might even go as far as banning the importation of ACTs in the future. The advantage to Nigeria of building up national ACTs production, as opposed to relying on imported products as it does now, even at subsidized prices, would be security of supply and the creation of an export industry with jobs in both the agricultural and manufacturing sectors.

Prospects for an ACTs Industry in Nigeria

Producing ACTs requires several different processes, both agricultural and industrial. It appears unlikely that in Nigeria a single company would try to take on all of the processes involved. However, because the interactions among several partners would be a relatively simple part of the enterprise, the hypothetical case study workshop held on April 24–25, 2006, decided that the hypothetical enterprise would be a single company that would incorporate all the necessary processes, from cultivation to manufacture. The participants concluded that any one of the component phases of the business could be operated at a profit: growing artemisia, growing plus extracting artemisinin, producing artemisinin derivatives, and manufacturing ACTs.

The workshop named the company Nigerian Anti-Malarials Ltd. Its objectives are to profitably produce ACTs to be sold in Nigeria at an affordable price through

- the cultivation of *A. annua* in Nigeria
- the local extraction and purification of artemisinin
- the production of artemisinin derivatives, possibly in collaboration with advanced laboratories

volunteers); (2) the team of assessors evaluates all the data presented, and, if satisfied with the evidence, sends the product to professional control testing laboratories contracted by WHO in France, South Africa, or Switzerland for analytical verification of quality; and (3) if the product is found to meet the specified requirements and the manufacturing site complies with GMPs, both the product linked to this manufacturing site and the company are added to a list hosted by WHO on a public web site.

- the local manufacture of ACTs that are globally competitive under cGMP-compliant conditions

Nigerian antimalarials would face competition from several very different sources, including other foreign and local companies growing artemisia elsewhere in Africa, other firms making and marketing ACTs in Nigeria, and companies using alternative technologies for producing ACTs or offering alternative products. If the Nigerian company does not qualify for subsidies and grants, the most serious competition would be from other companies benefiting from such funding, which would enable them to sell at a price below cost.

Aside from China and Vietnam, significant artemisia production is under way in Kenya and Tanzania, and start-ups in Senegal, Madagascar, Ghana, and Cameroon have stated their intentions to manufacture ACTs for the African market. Furthermore, a Chinese producer recently informed the minister of health of its intention to manufacture ACTs in Nigeria, using artemisinin from undisclosed local sources. Alternative technologies, including the synthetic or bioengineered manufacture of artemisinin, are in development, but they are not expected to appear on the market for at least 5–10 years. The most competitive alternative products are mainly the older, less effective (but less costly) malaria drugs. An educational campaign will be needed to convince the public of the superior performance of ACTs in reducing suffering and loss of life, as well as the cost benefits when the costs of hospital care and loss of productivity and wages are included.

On the positive side, Nigeria has the largest internal market for antimalarials in the world, estimated at 25 percent of the total global need. A company with WHO prequalification would have an opportunity to serve a large market in other countries of West Africa as well. The Nigerian pharmaceutical industry includes several companies that claim they have the capability to achieve prequalification for ACTs, although the fact that only one company in the world has so far succeeded demonstrates that the challenge is not trivial. As for the raw materials, recent trials with a variety of *A. annua* cultivars suggest that Nigeria has good growing conditions and may be able to produce multiple harvests annually with a good yield of artemisinin. The government has announced strong support for the local production of ACTs, and it appears to have the political will to provide concrete assistance. Success would open a niche market for Nigeria in the West African region and create jobs for the populace.

On the negative side, the inadequacy and unreliability of the infrastructure, including electricity, water, and roads, will be a problem. And there are few local suppliers of pharmaceutical supplies, which means that many materials will have to be imported, most specifically solvents for artemisinin extraction, which are currently unavailable on the local market. Another problem is the large market for non-ACT malaria remedies and the large supplier base for counterfeits and imitations. Furthermore, although the government has shown goodwill in addressing some of these problems for the ACTs market, there is a history of lack of implementation of government policies.

Growing *Artemisia annua*

Growing artemisia could be a profitable opportunity for farmers. It would enable them to generate income from a new cash crop with a guaranteed market, at least in the short to medium term. Because it is a medicinal plant, farmers producing it would be eligible to receive donated land from the state. Unlike coffee and tea, the major cash crops in the region, *A. annua* is an annual, and its acreage could be adjusted each year in response to the market.

Artemisia can be grown under a wide range of conditions. Internationally, relatively little standard agronomic research has been devoted to improving the artemisinin yield of *A. annua*. Research on selection and breeding might improve yields in tropical Nigeria and elsewhere. In China and Vietnam, the largest producers of *A. annua* at present, most *A. annua* is still collected in the wild. Trials of different cultivars of the plant have been carried out on a small scale in the humid Nigerian lowlands in the South-South region in Calabar, which has an annual cycle that avoids the dry season. Elsewhere, cultivation is begun immediately after the dry season in February and March, and then planting begins again in July and August. The use of drip irrigation may be necessary for production on a large industrial scale.

A senior researcher and his colleagues at the University of Calabar have undertaken considerable work to assess the artemisinin yield of various *A. annua* cultivars and to increase the efficiency of plant production. The capability of the Calabar group is a valuable resource as the nation contemplates the manufacture of effective malaria drugs. These researchers have led in the scientific testing of different artemisia varieties under different conditions to find the best cultivars for Nigeria, and they are in the final stage of negotiations for an industrial plant that could enable them to extract artemisinin from the leaves on a commercial basis.

Extraction and Purification of Artemisinin

The most commonly used process to extract artemisinin from artemisia leaves is solvent extraction, a well-known, reliable method. It is not vulnerable to electrical outages; it can be carried out in the tropics in open buildings with no walls; it operates at relatively low temperatures; it carries no risk of explosion; and it is not protected by patents. Solvents can be recovered and recycled, which lowers the cost of the process and helps to protect the environment.

The usual solvent is hexane, with an additive to protect against explosions. However, at present in Nigeria the solvent must be imported. If a new extraction facility is to be built for artemisinin production, turnkey plants are available from U.S. and European manufacturers, but the price is high and the scale, which is based on anticipated demand, is uncertain. Initially, it may cost about 15 percent more to grow, extract, purify, and derive artemisinin locally than to import artemisinin derivatives directly. To encourage local production, government or donor subsidies may be needed. Alternatively, the company could go into partnership with a foreign producer to share the cost of the extraction plant in Nigeria and send the material for purification and manufacture of the derivatives to an advanced laboratory. The Calabar group has received a preliminary offer from a Chinese company to import—and operate for six months—a plant to extract artemisinin and another small plant to produce artemisinin derivatives, all apparently below cost (i.e., subsidized by the Chinese company or government). The potential buyers for the products of extraction or the derivatives for the final manufacture of ACTs have not yet been identified.

Manufacture of ACTs

Several pharmaceutical companies in Nigeria are either presently marketing a combination therapy for malaria (consisting of two separate malaria drugs in blister packs) or stating their intention to manufacture ACTs. It would not be difficult for these companies to operate under the cGMP conditions certified by WHO, which would permit the product to be sold in Nigeria and other countries that are willing to accept the import. However, because the price of the product would be based on the full manufacturing cost, the market is likely to remain very limited. Without WHO prequalification, a company would not likely be eligible for international purchases using funds from the Global Fund, nor for a subsidy if that option materialized.

The cost of a facility to manufacture ACTs, which is high, is estimated in Appendix C. If a facility becomes a reality, it probably would be through an existing Nigerian pharmaceutical company, possibly one of those currently making blister packs or considering ACT production. Funds for additional equipment may be available from a commercial bank, an international finance agency, or a joint venture with a foreign company. The high probability of a government contract, at least for the population under age five, could make such a facility an attractive package. However much depends on the structure and application of the international subsidy for ACTs that is clearly in the works and the capability of the firm to achieve prequalification from WHO.

4

Meeting Needs

The issues raised in this report concern three entities: developing country governments that are unable to meet some of the basic needs of their populations, the philanthropic and international financial and foreign assistance agencies that are dedicated to assisting people, and entrepreneurs and businesses that may be able to address the unmet needs. Examples of the needs described here are electric energy for homes, safe water, and an effective malaria medication. These goods and services are not usually provided for free even by government. Most people expect to pay for them, but generally at subsidized prices that the majority can afford.

The proverbial life-sustaining needs—food, clothing, and shelter—are not addressed here, because few governments provide these necessities, even at subsidized prices. However, in the areas of energy, safe water, and medicines governments generally provide some services, because they affect not only quality of life and health, but also productivity, as any worker who has suffered from severe diarrhea or malaria can testify.

The reasons that some populations are not receiving power, water, or medications are usually financial and technical, not necessarily political. Rural populations may be far from the electricity grid and widely dispersed so that it might be uneconomic, even in wealthy countries, to provide electricity at a reasonable cost. The same is true of municipal piped water. Technical difficulties compound the problem; poor infrastructure causes intermittent and low-quality energy and water supplies, blackouts, and contaminated water, even in big cities. Effective malaria treatment, like the treatment for HIV/AIDS, is limited by the prohibitively high costs of the drugs compared with the earnings of most of the sufferers. It differs from HIV/AIDS, however, in that the latter is considered high priority by the United Nations and the developed world. A high-powered international effort is under way to provide sufferers with antiretroviral drugs, and many governments have provided significant funding. That said, malaria has received ever greater attention over the last eight years, and so resources may continue to grow. The drug distribution network for malaria is much different from that for HIV/AIDS, however, and requires different financing mechanisms.

Very often the task of remedying these deficiencies is left to the international and local nongovernmental organizations (NGOs) operating within the country. These groups are funded by philanthropic foundations, bilateral aid agencies, or international organizations. Their effectiveness is measured in terms of the number of homes or communities electrified, the water delivered or water purification systems installed, or the malaria treatments donated, but rarely are those most in need served by such efforts. Those who are recipients usually do not pay for the products or services, and so such solutions are rarely sustainable. When the funds are exhausted or the interest of the NGO or its donors changes, the service ends. People who have been receiving free power, water, or medicine find there is no infrastructure or commercial businesses to provide these goods and services at affordable prices, and often they are little inclined to pay substantial funds for what they had come to consider a free entitlement.

The strategy of private companies providing public goods and services, which is the subject of this report, has a role for governments, donors, NGOs, and, of course, private businesses. For governments, rather than attempting to provide commodities or services directly to needy communities and families, their policies should be aimed at encouraging private companies to fill these needs *while making a profit* (sometimes called "doing well by doing good"). The workshop reports outlined in this report and presented more fully in the appendixes describe how companies can operate at a profit in Nigeria while providing poor people with the three products and services used as examples. With profits to be made and where the banking and business environment is congenial, there is a good chance that the solutions will be self-sustainable and not require continual government input. This is particularly true in a vibrant entrepreneurial society like that of Nigeria if the solutions are properly managed and promoted.

MOBILIZING THE PRIVATE SECTOR TO PROVIDE PUBLIC GOODS

Although many companies profess to be motivated by a social conscience, fashions in social conscience sometimes change. For sustainability in providing services to the poor in developing countries, it is necessary to enlist the existing private, profit-making businesses and to encourage the creation of new ones. Incentives should be aimed at responsible companies that are willing to compete on price, quality, and service to the consumer. Three motivational factors would attract the kind of companies desired:

1. good probability of making a profit over the short to medium term at an acceptable risk, including legal risk
2. availability of credit for building the business, and, in some cases, for the consumer
3. availability of the skilled labor needed to run the business

An antimotivational factor for responsible companies would be an arrangement in which the government contracts directly with companies for products and services to be delivered to consumers or laws that require compliance by consumers, such as displays of solar collectors on roofs.

The first motivational point is most likely satisfied by the huge potential market of households who are served intermittently or not at all by the electricity grid, who do not have safe drinking water, or who suffer from frequent bouts of malaria. However, most families will not have access to cash to purchase a solar home electrical system. Nor will a community have the resources to order a community water system. Nor is there easy access to consumer credit based on the system itself as collateral. (The situation is somewhat different for artemisinin, which is discussed separately later in this chapter.) Furthermore, many potential consumers are not aware that some of the new, affordable technologies will allow them to improve the quality of their lives and health.

An argument for private sector involvement is indeed that many different technologies are available for exploiting a renewable energy resource or purifying water at low cost. For electric power, these technologies include several models of solar home systems, microhydro systems for those living near streams, and small wind systems for those in windy areas. For safe water, many technologies besides the two discussed in the workshop—ceramic filters and ultraviolent (UV) purification—are available. Because there may be reasons why a particular design is inappropriate for a particular family or village, incentives to companies about the choice of a particular product or system should be flexible. The inclusion of a maintenance contract, preferably paid periodically rather than up front, would seem to be required for all poor families. It probably would also be required by a supplier of consumer credit, who would test compliance regularly, at least for community-size water purifiers and solar systems. The governments of some countries offer coupons that can be exchanged for payment for service contracts on solar energy systems. But this scheme should be considered with care. Because banks are likely to require a service contract, coupons might lead to a situation in which government intervention to provide coupons becomes a part of the consumer credit system. An alternative method is to include the service contract as part of the cost of installation, with incremental payments made on the contract through the term of the loan.

For solar electric systems, availability of consumer credit would be an important incentive to the would-be installer. In India, SELCO itself generated the market for consumer credit by means of an intensive training and education campaign for bank officers until nearly all banks in the state agreed to lend for solar systems (see Chapter 3). Alternatively, the government might guarantee such loans under the appropriate conditions. In the United States and other countries, tax incentives are offered to the purchasers of renewable energy systems.

A government-sponsored public education campaign could alert consumers to the advantages of light in the home and safe water for children, and at the same time alert companies to the possibilities of supplying these products. Appropriate slogans might include: Wash your hands! Filter your water! Light up Nigeria with renewable energy! Guaranteed credit available for home solar lighting systems!

Consumers will need training to use many of these systems in the home. Training will generally be the task of the sales agent, but the government might also assist. Experience in other countries with properly trained home extension agents in rural areas demonstrated that they could detect health risks and encourage solutions, while training families in the need for and use of filtered water or home solar systems.

Legislation encouraging the use of renewable energy sources, in general, perhaps with tax benefits, would encourage the use of solar home systems. However, there is an urgent need to clarify the liability of a private company distributing water or electrical energy to homes and communities. At present, they are at risk of prosecution as intruders into the prerogatives of the related ministry, and no new industry will be created while that situation exists.

The role of the government from this perspective is that of enabler, not implementer. This philosophy is applied in many countries, developed as well as developing countries, in the public works arena. For example, private companies build toll roads on public land; they collect the tolls for a certain number of years to reimburse them for the costs and to make a profit; and then they turn the roads over to the government. These associations are generally governed by an initial contract and continuing regulation. For the three examples proposed in this report in which the products and services are delivered to individual families, the situation could be simpler. Dozens of separate companies would provide the products and services, but, as enabler, the government would make any agreement with the customer, who would take responsibility for selecting the vendor and the product, or with a much smaller number of financial institutions, which would monitor compliance.

THE SPECIAL CASE OF ARTEMISININ

It appears to the committee for this study that the potential for success of an initiative to manufacture artemisinin combination therapies (ACTs) for the local market, the largest in the world, and for export depends on adequate global financing for the drugs. The strategy under development for financing is a global subsidy, first proposed by the U.S. Institute of Medicine, which is part of the U.S. National Academies (see Chapter 3).

Every interested country has a national committee that deals with the Global Fund (see Chapter 3). In Nigeria, this committee or some other coordinating body should become an active player and prepare to make the first proposal. Such a step may require contact not only with the World Health Organization (WHO), but also with the potential funders directly. It is important for all players, growers, and pharmaceuticals, to work together, in coordination with the government, to put Nigeria into the international ACTs business. Strong leadership is required at the national level. The economics and financing of such a venture deserve considerable analysis and scrutiny, not only by the private sector actors that might be involved, but also by government, possibly academic researchers, and possibly the Nigerian Academy of Science. Understanding the full economic implications of starting ACT production will be vital to a sound enterprise.

The international standards for product quality and appropriateness are yet another aspect of artemisinin manufacture that must be studied carefully. These standards, WHO's current Good Manufacturing Practices (cGMPs) and prequalification, were described in Chapter 3. Fully accredited manufacturing will be essential for Nigerian products to be eligible for international financing (e.g., through the Global Fund) or for a global subsidy, should one become a reality.

There are many ways in which the government could promote the manufacture and distribution of ACTs. Direct actions could begin by providing suitable land for growing artemisia without preconditions on employment or distribution that might inhibit the agility of the proposed company to make a profit. This approach might be followed by a program to provide first-stage financing for new start-ups or grants to encourage pharmaceutical companies to enter the market. Assistance to companies to achieve prequalification would shorten the timetable. This assistance could be reinforced by using the purchasing power of the government to make an advanced market commitment in order to serve the children under five and any other populations to whom the drugs have been promised at government expense. However, donations of free medicine will reduce the market and might adversely affect companies that expect to make a profit by providing Nigerians with ACTs, unless the product was procured locally.

A secondary level of assistance would involve standards, quality control, and steps to prevent counterfeits. Counterfeit ACTs have already shown up on the international market with serious consequences for patients.[1] All counterfeits, clones, and low-quality products, including non-ACTs, should be banned from the Nigerian market to protect the public and assist the new manufacturers at the same time. This effort should be complemented by public campaigns encouraging Nigerians to use bed nets, clear away standing water, spray walls with DDT, and take the correct medicine (ACTs rather than monotherapy or blister packs). Materials and strategic plans are available from WHO and other agencies. Finally, a program of public education in the schools on the avoidance and control of malaria should be cost-effective and successful, as similar programs on HIV/AIDS and smoking have been in other places.

The government could also take the following specific actions to encourage ACT production:

- providing waivers on duties on imported equipment, raw materials, solvents, and other materials involved in ACT production
- enforcing a ban on artemisinin monotherapies
- assisting in training and research on the agronomy of artemisia and best ACT formulations for Nigeria
- arranging media advertisements to promote effective malaria therapy and completion of courses of treatment
- passing legislation to require the distribution and use of ACTs at all government clinics, hospitals, and agencies

[1] *PLOS Medicine* 3(June 2006), Public Library of Science, http://medicine.plosjournals.org/archive/1549-1676/3/6/pdf/10.1371_journal.pmed.0030197-S.pdf.

With government assistance, the Nigerian malaria community must inform itself about the state of play on the international scene, which is critical to entering the international market. The world has recognized that malaria requires a special effort. New coformulations are in the late stages of development by various groups, including the Drugs for Neglected Diseases initiative (DNDi) and the Medicines for Malaria Venture. The next steps involve developing licensing agreements with manufacturers to produce the drugs. Nigeria must constantly be aware of progress and contribute to the process as opportunities arise. There will be no other way to succeed in this market.

LINKING SCIENCE-BASED ENTERPRISES TO RESEARCH

The proposals advanced here are ambitious. If 100 million people in Nigeria are without safe water, home lighting, or effective malaria therapy, then tens of millions of households may be a potential market. Companies such as SELCO and WaterHealth International serve tens of thousands in the countries in which they are active, and a thousand similar enterprises could be required to meet the needs of all Nigerians. The demand for imported solar cells, UV filter units, and solvent extraction plants for artemisinin would have a large impact on the present world market for these products, and Nigerian companies and local subsidiaries of international companies would emerge to capture some of that market. Because the science-based solutions in this study are complex technological products for which global competition has not been great thus far and which are not presently manufactured or used in Nigeria, there will be an ample opportunity—and even necessity—for innovation, if only to avoid patents. New jobs will appear for engineers, scientists, and public health workers, among others, and universities will be presented with new opportunities to undertake both research and training.

Some Nigerian university laboratories are already active in these areas. And other universities in Nigeria will find it advantageous to enter these and other fields with similar potential, such as low-cost construction, solar cooking and refrigeration, and micro wind and hydroelectric energy sources. Many would benefit from scientific cooperation with universities overseas, and should consider innovative approaches to that process as well.

Traditionally, international scientific cooperation between universities has been dominated by the exchange of faculty and students for training and research. In the last few years, however, international travel has in some ways become more difficult; it is now harder to obtain visas, and developed country universities have fewer resources to accommodate foreign visitors. Moreover, the costs per visiting scientist and student have risen; research materials have become more expensive; accounting of materials and equipment has become more rigorous; and grants are less flexible to accommodate visiting scientists and students.

Meanwhile, technological advances such as information technology have made the traditional modes of cooperation less necessary. For example, by means of long-distance learning students can take classes online, and effective collaboration among researchers can be achieved via the Internet, requiring few visits between laboratories. Video teleconferencing is another useful means of linking researchers internationally. The tools of information technology should be deployed to optimize collaboration and productivity in the research effort that will accompany the strategy proposed here.

5

Recommendations

INCENTIVES FOR PRIVATE COMPANIES TO PROVIDE PUBLIC GOODS AND SERVICES

The proverbial 100 million in Nigeria lacking home lighting, safe drinking water, and effective malaria therapy is a population larger than those of all but a handful of countries. Lacking the resources to more than double the number served by the electricity grid and piped water network, the Nigerian government can turn to the very active private sector to provide these basic products and services. A program of financial incentives and technical assistance to encourage responsible service will be in consonance with the National Economic Empowerment and Development Strategy (NEEDS) poverty reduction plan prepared with the International Monetary Fund (IMF). If a large proportion of those lacking the basic services of electric power and safe water in the home are to be served, a solution engaging the private sector may require the involvement of a large number of new companies, many of them start-ups. First-stage financing or venture capital will be required, perhaps for the majority of them.

With some trepidation, we mention a mechanism—the Alternative Agricultural Research and Commercialization Corporation (AARC)—that was set up by the U.S. Department of Agriculture (USDA) in 1992 to provide venture capital for the promotion of alternative systems of agriculture in the United States. The AARC, which had an independent board of directors, made investments in mostly small companies that promised to commercialize agriculture-based industrial products. It took an equity position in the company or a royalty on sales, and it had an agreed exit strategy after eight years. Unfortunately, the AARC was disbanded after seven years when the USDA Inspector General's Office uncovered serious deficiencies in its management and oversight.[1] Despite this outcome, a similar mechanism with a narrower mandate, perhaps run by a nongovernmental organization (NGO) that could be held to account, might prove effective in funneling venture capital funds to small companies that would directly further the aims of the program. An essential ingredient of any such program is effective evaluation and approval of the product and associated service for quality and compliance with published standards as well as the oversight that was ineffective in the case of AARC. Like many venture capital firms, the proposed one could also offer technical assistance with logistics and help in negotiating the legal requirements.

Alternatively, a large major company in the solar energy or pure water business could be enticed to develop a system for sales in Nigeria and offer franchises to entrepreneurs to sell and service units in different parts of the country, providing credit and training as required.

In the previous chapters and in the following recommendations, reference is frequently made to the "Nigerian government." However, in Nigeria, as in most countries, government exists at several levels—generically, federal, state, and local. The present distribution of power frequently allows the state and local levels of government to undertake actions such as those proposed here. Moreover, they may have easy accessibility to the poor rural populace and community-based projects that are involved. Here we do not try to determine which level of the Nigerian government system is most appropriate for the indicated action. The term *government* should therefore be understood to apply to all three, with the hope that the recommendation will be considered by the one that is most interested and effective.

Recommendation: **The Nigerian government should develop a system of incentives to encourage private companies to sell and service solar electric systems for the home to rural residents who are not connected to the national grid. The Nigerian government also should develop a system of incentives to encourage private companies to sell and service water purification systems to communities that are not served by municipal or national water supplies, or to produce household filtering systems for safe drinking water.**

---

[1]U.S. Department of Agriculture, "Assessment of the Alternative Agricultural Research and Commercialization Corporation—Management Lacking over High Risk Investments," 1999, http://www.usda.gov/oig/webdocs/37099-1-FM.pdf.

Consumers of home solar electric systems will need consumer credit to purchase their units. Local financial institutions, community associations, cooperatives, and trade associations should be encouraged and trained to make small loans for such systems, including service contracts, secured only by the system itself, as has been done in other countries. The experience of SELCO in India demonstrates how the company itself can intervene effectively, but if a large number of new companies will have similar needs, the government could take the lead in working to establish a partnership with the credit organizations.

Recommendation: **The Nigerian government should work with banks, community associations, cooperatives, trade associations, and other financial institutions to establish microloan funds that would be dedicated to providing consumer credit for home solar electric and water filtration systems in rural areas.**

Many householders are not aware of the advantages to their families and the improvements in their lifestyles and educational and business opportunities that are made possible with electricity. Many also do not realize the harmful effects that contaminated water can have on their children and the cost to themselves and the nation. Mass media campaigns could be highly effective in promoting these lifestyle benefits. Suitable slogans might be along the lines of "Wash your hands, filter your water," or "Light up their lives" (with children shown studying beside an electric light).

Recommendation: **The Nigerian government should sponsor an educational campaign to encourage people to invest in electric power and safe water.**

Existing law might actually make it legally hazardous for private companies to bring electric power into people's homes, because this service is now the exclusive mandate of a federal ministry. A new law in the realm of public health and conservation of natural resources could encourage people to conserve water and energy and to avoid disease from infected water and food, in part by making it legal to install electric power and water filtration systems and sell and deliver electric power and safe water to the home.

Recommendation: **A legal remedy should be found to the prohibition on private companies or individuals providing power to homes**

in Nigeria. In the context of legislation on renewable energy sources, it might give additional impetus to the proposed program.

THE CHALLENGE OF ARTEMISININ COMBINATION THERAPIES

The production of artemisinin combination therapies (ACTs) by Nigerian companies, even just for the Nigerian market, will probably not be successful without active and purposeful government intervention. Currently, economics is holding back the demand for ACTs. The strategy that has received the most attention by those trying to solve the problem is a global subsidy, which would be applied at the supranational level and would be financed in large part by international donors (see Chapter 3). Only high-quality and appropriate products (probably prequalified by the World Health Organization) would qualify for the subsidy. Should the notion of a subsidy not succeed, it is unclear whether the majority of those seeking malaria treatment will ever have access to these drugs. Today, however, nothing about the proposed subsidy is certain.

Nigerian pharmaceutical companies and public health officials may not be sufficiently aware of developments in the international arena that will affect their plans. The government should play an active role as an information broker—with well-informed international contracting assistance as necessary—to ensure that all interested officials and companies are fully briefed. At the same time, those preparing to cultivate artemisia or to take other steps toward producing ACTs for the Nigerian and world market should be encouraged and assisted.

Whether or not *Artemisia annua* is grown commercially in Nigeria and Nigerian pharmaceutical companies are able to participate fully in a subsidized distribution system, actions must be taken to ensure that the population is as fully protected as possible. These actions include public campaigns calling for the use of treated bed nets, avoiding counterfeit and ineffective drugs, and draining potential breeding places for mosquitoes.

Recommendations:

The Nigerian government should support private sector participation in the global ACTs market. It should do this by tracking international developments related to the economic and technical requirements of all aspects of ACTs production and establishing formal and informal links to academic, public, and for-profit entities that could play a role in ACT production. The Nigerian Academy of Science could play an important role by convening potential

participants for the exchange of information with the government and the international community.

The Nigerian government should ban all counterfeit drugs, illegal clones, low-quality products, and artemisinin monotherapies from the Nigerian market by means of proactive enforcement against illegal activity. It also should prohibit all advertising for such products.

Duties and other impediments to the importation of the equipment, raw materials, solvents, and other materials needed for ACTs research and production should be removed. The Nigerian government should support training and research on the agronomy and selection of the best cultivars of *Artemisia annua* in collaboration with the global malaria community. Public health laboratories should participate in surveillance to determine the levels of resistance to possible partner drugs for different ACTs formulations.

CONSUMER EDUCATION AND TRAINING

In every modern society, consumers are confronted by a bewildering array of choices of products and services that affect their health, well-being, and economic security. However, most countries offer little formal training, and so consumers are expected to educate themselves by means of the media, their friends and community, and commercial advertising. Sometimes, consumers receive some reassurance from the public regulatory agencies or the private NGOs that monitor product safety, drug efficacy, and truth in packaging. But when these watchdog agencies are ineffective or absent, or when populations are illiterate or isolated, public agencies must step in and provide more explicit consumer education through the schools or in public campaigns or pronouncements. A prominent example is the series of pronouncements by the U.S. surgeon general on the dangers of smoking, which must be reproduced on all cigarette advertisements in the United States.

An idea that has been put into practice successfully in Africa is a university-to-village extension program. Engineering, nursing, medical, and other professional students "adopt" villages to assist. In doing so, they make frequent visits and provide free service and advice to the villagers on simple solutions to common problems. At the same time, they become sensitive to the villagers' problems and issues, which serves to make them better professionals and better future citizens. The hands-on work can be integrated into the curriculum to formalize the training and recognize the educational benefits to the students. Such an approach was taken in the introduction of the simple filtration of water in Bangladesh.[2] The Kigali Institute of Technology (KIST) in Kigali, Rwanda, has been using such an "adopt a village" model in its engineering program.[3] Students are exposed to villagers' problems at an early stage in their curriculum.

Recommendation: **The high mortality rates in Nigeria from diarrheal disease and malaria argue that the Nigerian government should offer health education and training in the schools that would include the importance of safe drinking water, how to maintain a sanitary water supply, and how to choose effective medicines. Other relevant topics could include nutrition, hygiene, safe sex, and the prevention and treatment of common diseases such as respiratory infections, diarrhea, and HIV/AIDS. This program should be supplemented by public pronouncements on topics such as filtering water and a public information campaign on the importance of using the most effective antimalarial therapy—currently ACTs—and the necessity of completing the course of treatment.**

THE ROLE OF PHILANTHROPIC FOUNDATIONS AND DONOR AGENCIES

The proposals offered here for government suggest a new approach for philanthropic foundations and donor agencies. Ideally, any donor should have two objectives. First, enable the target community to find and implement solutions to its own problems rather than giving it a particular solution—that is, as the popular saying goes, teach people to fish instead of giving them fish. Second, create sustainable solutions that do not require an ongoing supply of donor funds.

Too often, however, aid agencies and foundations are criticized for making recipient countries and communities dependent on aid by giving them unearned benefits that are not sustainable or by creating wealth that is not shared by the majority of people. Furthermore, the benefits disappear when the aid program ends because the local government is unable to continue the support.

Instead, we suggest that this kind of philanthropy should be considered a sort of investment in public goods and services. The goal should

---

[2]Young-Gun Zo, Irma N. G. Rivera, Estelle Russek-Cohen, M. Sirajul Islam, A. K. Siddique, M. Yunus, R. Bradley Sack, Anwar Huq, and Rita R. Colwell, "Genomic Profiles of Clinical and Environmental Isolates of Vibrio cholerae O1 in Cholera-Endemic Areas of Bangladesh," *Proceedings of the National Academy of Sciences* 99(2002):12409–12414.

[3]See http://www.kist.ac.rw.

be to create enterprises that will work to continue to provide the goods and services that represent the solutions to socioeconomic problems. If necessary, such enterprises could be asked to return a portion of their profits to the donor or to a fund that will expand the activity. They will be sustainable, because the owners and employees themselves depend on the success of the enterprise—rather than renewal of grants—for their survival.

However, for donors this path is the more difficult one, because the creation of successful enterprises is less well understood in the philanthropic world than grant giving, and a natural failure rate characterizes even the most fertile investment plans. Donors would do better to emulate the most successful venture capital firms in Silicon Valley by employing the kind of expertise needed to select the best candidates and offer the kind of technical advice needed by recipients. Even so, the effort should be spread among many enterprises to improve the probability of success, and thus it will resemble a balanced investment portfolio more than a normal grant program. It might be initiated with the kind of knowledge assessment that was employed for this project to create a model business plan, with the implied hope that some of the selected participants themselves will be among the first applicants for start-up loans, or technical assistance, or whatever the program offers as an incentive.

Some NGOs, which often serve as agents of the donor agencies, also should modify their method of operations. They are adept at organizing demonstration projects, some of which might illustrate the feasibility of installing solar PV systems in villages by donating such systems and teaching the recipients to use them. Experience has shown, however, that, relieved of the responsibility of selecting and paying for the unit, the recipient has little motivation to maintain or use it effectively, and the project often ends up demonstrating to villagers that such utilities need not be paid for and have a limited lifetime. Such an approach is detrimental to entrepreneurs who seek to sell and service such units, and should not be part of the proposed effort.

Demonstration projects would still be useful, but the demonstration should be of small enterprises that make a profit by supplying the basic needs of poor people. Experienced businesspeople, local or international, could be enticed to assist these companies, even to invest in them, and, even further, to make a second career as a consultant helping other companies to enter these markets. As Professor C. K. Prahalad has pointed out, real profits are to be made by serving the people at the bottom of the pyramid, but the business plan is different (see Chapter 1). There is much to be learned.

Another area in which NGOs excel is public education and media promotion. With the support of their donors, they could take the lead in the campaigns for safe water, home solar electric systems, and the exclusive use of ACTs for malaria treatment, and they could serve as effective partners of the enterprises that provide the products and services.

Meanwhile, the many donors and NGOs that already employ similar techniques with success should be supported and encouraged. The philosophy of philanthropy proposed here raises some associated problems that should be solved. In the United States, philanthropic foundations must register with the Internal Revenue Service for tax exemption under Section 501(c)(3) of the Internal Revenue Code. One of the conditions is that they not make grants to profit-making enterprises, even with charitable intent. Many such grants are made legally, however (for example, reportedly the Rockefeller Foundation made a start-up grant to SELCO in India), whereas other companies continue to confront obstacles in trying to secure such support. Each foundation must evaluate its situation in each case and find an appropriate way to support these enterprises. The same problem may arise when the subsidy for ACTs is finally established and private companies are paid above-market prices for their medicines. This committee does not find itself competent to assist with these problems.

> *Recommendation:* **Philanthropic foundations and donor agencies should orient some of their activities in developing countries toward creating and supporting profit-making enterprises that would provide public-benefit goods and services to poor people. Grants should be replaced in spirit with first-stage financing or investments, and the portfolio should be broad enough in diversity of enterprises with different business plans and different technologies to raise the probability of financial success in this area, where there is relatively little experience.**

THE ROLE OF THE NIGERIAN ACADEMY OF SCIENCE

The Nigerian Academy of Science, like the U.S. National Academies, selects its own members on the basis of scientific merit and contributions to knowledge. Within Nigerian society, it represents the national scientific community, and its officers often participate in national events in that role. It sponsors scientific conferences, awards prizes, and publishes a national scientific journal.

The U.S. National Academies are distinguished among the world's science academies in their formal role as advisers to the nation on science, engineering, and medicine. Their original mandate as self-selecting scientific societies, issued by the U.S. Congress in 1863, included a mandate to advise the government in matters of science and art upon request, but added that the Academies "shall receive no compensation whatever for any services to the Government of the United States." They therefore receive no appropriation from Congress, but they publish about 250 study reports a year, 90 percent of them commissioned and paid for by the U.S. government. The absence of compensation refers to the diverse volunteer scientists and engineers who, as independent members of the study committees, research and write the reports, or as anonymous reviewers provide independence and quality control. In few other countries does the national science academy play as important a role in public life.

The Nigerian Academy is well positioned to strengthen its public role, and several actions related to this report might help it to move in that direction while furthering the goals of sustainable development.

A good framework for a proactive role by the Nigerian Academy of Science could be modeled on a program that has been a fixture at the U.S. National Academies since 1984. The formal mission of the Government-University-Industry Research Roundtable (GUIRR) is "to convene senior-level representatives from government, universities, and industry to define and explore critical issues related to the national and global science and technology research agenda that are of shared interest; to frame the next critical questions stemming from current debate and analysis; and to incubate activities of on-going value to the stakeholders. This forum is designed to facilitate candid dialogue among participants, to foster self-implementing activities, and, where appropriate, to carry awareness of consequences to the wider public."[4] The chair of GUIRR is usually a retired high government or industry official, and heads of major government departments are ex officio and entitled to appoint representatives. Other members must agree to participate in person. Reports of GUIRR workshops are often published for public information. Recent examples include: *National Laboratories and Universities: Building New Ways to Work Together* (2005); *Envisioning a 21st Century Science and Engineering Workforce for the United States: Tasks for University, Industry, and Government* (2003); and *Overcoming Barriers to Collaborative Research: Report of a Workshop* (1999).

In the Nigerian context, the debate need not be limited to research, but could embrace other areas in which the governmental, university, and industrial sectors have a common interest. The first topics for discussion and perhaps publication could be how to mobilize the private sector to sustainably provide basic services to remote parts of Nigeria; setting and enforcing government standards for drinking water sold to the public; and a national strategy for ensuring that Nigeria has a secure supply of effective malaria therapy at affordable prices. Another valuable topic for discussion could be preventing the illegal sale of counterfeit products and medicines.

---

[4]U.S. National Academies, http://www7.nationalacademies.org/guirr/index.html.

Recommendations:

The Nigerian Academy of Science should establish a program similar to the Government-University-Industry Research Roundtable (GUIRR), focusing on scientific and technological issues of common concern. It should sponsor and convene workshops of experts drawn from the government, academia, and industry, and reports of the workshops should be published by the Academy.

One early topic for a meeting of the Nigerian roundtable and a workshop of experts should be the production of ACTs in Nigeria. Invitees would include agriculturalists, pharmaceutical companies, and government health officials. The workshop would seek to help interested companies to become informed about international efforts to subsidize the price of ACTs and the requirements for cGMPs (current Good Manufacturing Practices) certification and World Health Organization (WHO) prequalification to manufacture ACTs.

Another, equally important early topic for the Nigerian roundtable should be the importance of safe potable water to public health in Nigeria. A follow-on workshop that includes experts from academia, government, NGOs, and the private sector would bring the issue to the public. Especially useful participants would be representatives of the Ministry of Health, the Federal Institute of Industrial Research, the International Center for Business Research, the Nigerian Agency for Science and Engineering Infrastructure, and the Nigerian Association of Small and Medium Enterprises. Importantly, such a workshop could urge the government both to clarify the law on the right to provide potable water to households and to mount a campaign in favor of filtered purified water to combat diarrheal disease.

The U.S. National Academies should be prepared to assist the Nigerian Academy of Science to organize the Nigerian roundtable, if requested, in view of the National Academies' long experience with GUIRR. The National Academies also might assist by arranging for international experts to participate in Nigerian Academy–sponsored workshops dealing with solar energy, safe water, and ACTs.

U.S. scientific agencies with international programs, such as the Office of International Science and Engineering of the National Science Foundation and the Fogarty Center of the National Institutes of Health, should guide the exchange programs between the United States and countries such as Nigeria toward cooperation in helping small and medium enterprises to provide public goods and services. Many U.S. scientists have valuable expertise in linking research to enterprise creation.

Appendixes: Workshop Reports

The three workshops described in the appendixes that follow were carried out as part of the Science-Based Enterprise Creation project, a collaboration of the U.S. National Academies and the Nigerian Academy of Science.

The workshop on solar photovoltaics described in Appendix A was held December 8–9, 2005, at the offices of the Nigerian Academy of Science in Lagos. Appendix B presents the findings of the workshop on water purification, held December 12–13, 2005, at the same location. Appendix C reports the findings of the workshop on artemisinin-based malaria therapy. This workshop was held April 24–25, 2006, at the offices of the French Academy of Sciences in Paris, following a preliminary meeting at the International Institute of Tropical Agriculture in Ibadan, Nigeria, December 5–6, 2005.

The format of the workshops was based on the U.S. National Academies' knowledge assessment methodology, prepared in collaboration with the World Bank. This methodology is used to probe and explore the advantages and challenges facing specific science-based enterprises in a host developing country. The particular technique employed is the hypothetical case study. This exercise helps to identify the opportunities and challenges confronting enterprises seeking to exploit specific technologies for profit. It requires collaboration between foreign experts familiar with the technology and enterprise building and local experts familiar with the host country's social and economic situation.

During the workshops, the Nigerian experts played the role of trust-ees of a new company formed to exploit the technology, and the foreign experts acted as consultants, well versed in the problems surrounding the technology studied, but not familiar with the Nigerian business environment. The interaction of the two groups of experts was used to produce a plan for the new (hypothetical) enterprise that points out the opportunities presented by the new technology but also the pitfalls and challenges that must be confronted in Nigeria.

This report thus serves an audience interested in developing this particular technology in Nigeria, but it also is of interest to those concerned with the creation of knowledge-based enterprises in Nigeria in general and those concerned with exploiting this technology in other countries. This report is almost entirely drawn from the expertise and experience of the participants in the workshop and is not based on any independent technical or market research. By means of this report, the U.S. National Academies do not intend to either propose or discourage real investments in actual enterprises similar to those described here for Nigeria.

Appendix A

Solar Photovoltaics: Hypothetical Case Study

This hypothetical case study on solar photovoltaics was intended to explore the opportunities and challenges inherent in investment in a new enterprise that would capitalize on the effectiveness and low price of solar technology at the home or village level by producing, installing, and maintaining the solar units and accessories in rural villages. At present, no actual company is engaged in all these activities, although some companies, a few of which were represented at the workshop, market solar units to hotels and other higher-demand users.

Participants included representatives of the National Agency for Science and Engineering Infrastructure, Nigerian Association of Small and Medium Enterprises, Cross Rivers State University of Technology, Obafemi Awolowo University, University of Lagos, University of Nigeria at Nsukka, University of Ibadan, and the private sector. The U.S. National Academies were represented by Walter Ratterman, energy consultant with SunEnergy Power Corporation, Oregon, United States; M. R. Pai, executive director, Solar Electric Light Company, Bangalore, India; Wole Soboyejo, Princeton University and a member of the National Research Council committee for the project; and Michael Greene, project staff director at the National Research Council. The Nigerian Academy of Science was represented by its president Gabriel Ogunmola, and by its vice president, Njidda Gadzama. A complete list of participants appears at the end of this appendix.

SOLAR PHOTOVOLTAIC TECHNOLOGY

Photovoltaics (PV) is a way of converting sunlight directly into electrical energy. Photons in sunlight excite electrons from the valence bands into the conduction bands of a semiconductor, enabling the generation of electric current. This photoelectric effect was discovered in the nineteenth century, and the theoretical explanation won a Nobel Prize for Albert Einstein. The cost of early units was high because of the expense of producing high-quality silicon ingots and manufacturing the silicon wafer that is at the heart of this technology. As the price of silicon wafers has fallen with improvements in the manufacturing technology, the efficiency of the commercial units has increased to 15 percent with the use of crystalline silicon. Amorphous silicon wafers are cheaper to manufacture, but they are less durable than crystalline wafers, and they have a 5–10 percent lower efficiency that would require a larger solar panel for the same power.

Solar photovoltaic systems suitable for rural households usually consist of several components. They include a PV module containing the silicon cells to be mounted on the roof or another sunny spot, a battery for storing electrical energy for use at night, a charge controller, wires and structural frames, and outlets for lights and other appliances. Such a system can operate several fluorescent lamps (often four), a radio, a black and white television, and perhaps a fan. The system normally operates on 12 volts DC. Long-lasting deep-cycle batteries, which can discharge 80 percent of their charge during extended overcast weather, are best, but automobile batteries, commonly available in Nigeria, may be used as well. The charge controller prevents damage to the system in case of overcharging by the solar module or prolonged battery discharge from overuse. The cost for a 40–peak watt system is about $350–$500 worldwide, depending largely on the input duties on the solar panel. Other requirements are the cost of installation, periodic battery replacement (once every five years), and training for the user, all of which are often part of a service contract for maintenance. But without special arrangements, that price is out of the reach of most Nigerians in a country in which the annual per capita income is about $250.

APPLICATION EXPERIENCE

One of the more successful enterprises selling, installing, and servicing solar home systems (SHSs) is the Solar Electric Light Company of India (SELCO). SELCO Solar Light Pvt. Ltd., with a registered office in Bangalore, Karnataka, was founded in 1995 with initial financing from the Rockefeller Foundation. It is the first rural solar company in India to be engaged wholly in designing, marketing, and servicing a wide range of solar-powered equipment and installations for lighting, TV and radio, water pumping and purification, and many other applications. For the first two years of the company's existence, however, solar electrification was little known, and there were no sales. Initially, the company installed solar units in prominent places such as the houses of local village chiefs and local religious buildings. These systems acted as demonstrations to other villagers and local financial institutions.

From its earliest years, SELCO realized the importance of consumer financing, and it spent much of its human and financial resources to train bankers in the usefulness of the solar technology. SELCO conducted several rural bank sensitization programs, such as training for bankers on technology assessment, demonstrations at bank premises, and bank-customer meetings. As a result, the confidence of bankers in financing SHSs steadily increased. Within five years, SELCO was able to train and convince more than 550 managers of seven different local banks of the value of financing solar home lighting systems.

In 2002 a United Nations Environment Programme (UNEP) electrification project in India began to create a steady market for SHSs, and interest rates fell. In 2002 household systems cost $600, the same as a motorbike, but with market opening and competition building, the costs at present are about $400. SELCO now has 50,000 customers in the state of Karnataka, and agreements to finance SHSs with all the banks in the state, which require no collateral except the system itself. Customers borrow at market rates of $400 for a 40-watt system, repaid in five years. Alternatively, SELCO offers a "lease to own" scheme in which the consumer pays 10–25 percent of the cost of the system as an upfront payment, and the rest is lent to him or her by a rural bank at the priority sector rate of 12–14 percent interest per annum on a three- to five-year tenure. One of SELCO's partners is the Karnataka Vikas Bank (previously known as the Malaprabha Grameen Bank), a rural development bank with one of the highest recovery rates in the country. With its 200 branches in the Dharwad and Belgaum districts of Karnataka, the Karnataka Vikas Bank is known for its innovative microcredit schemes.

SELCO procures systems from reputed manufacturers after securing factory guarantees of quality, which are passed on to the consumer as performance guarantees. It has set up branches in villages, and its teams of local technicians on motorcycles ensure quick after-sales service and regular collection of loan installments. A typical SELCO branch has its own set of technician-salesmen and collection agents, all hired locally. The technicians work for SELCO on a salary and commission basis, which acts as an incentive to sell more systems. The company now has 170 employees in 25 centers and over $3 million in revenues a year. SELCO assumes total responsibility for (1) organizing awareness campaigns in rural areas, including PV system demonstrations; (2) identifying and prequalifying potential beneficiaries; (3) training local technicians, installers, and service personnel; (4) installing solar home systems purchased through the lending bank; (5) educating users; and (6) providing after-sales service and maintenance.

The SELCO experience has established that the distribution of PV systems via commercial channels without dependence on government subsidies can be a profitable business in rural areas of India. The following factors have contributed to the success of the SELCO model:

- By collaborating with local existing financial institutions, SELCO was able to gain the confidence of the people, which made the task of installation and collection of payment easier. Through these credit schemes, SELCO has been able to get around the high capital cost problem of PV devices, which is normally the biggest obstacle in rural areas.
- By setting up rural branches and training local people as technicians, SELCO has been able to generate local support as well as employment opportunities.
- By creating a local infrastructure for repair, SELCO has been able to provide quick and effective service in case of any faults with the systems. All travel is by motorcycle.
- By establishing local branches and operating them as independent entities, SELCO has been able to keep the management system decentralized and simple. A custom home lighting system can be designed and installed within 24 hours.
- By offering commissions to technicians, SELCO has ensured that they have a direct stake in promotion of solar technologies.

SELCO's partners have been nongovernmental organizations (NGOs), microfinance institutions, banks, cooperatives, and its rural customers. NGOs have been particularly valuable. They are active in the villages, promoting the solar photovoltaic systems, and they have helped to manage and maintain the systems. They were also instrumental in convincing the banks and cooperatives to support SELCO.

The largest competitor for solar home systems in Karnataka state is a division of Shell Oil, but SELCO has the largest market share. A growing market for SELCO is the urban and rural street hawkers, who typically work from 5:00 to 9:00 in the evening selling food and vegetables. Their daily earnings are about $2. Most of these hawkers use kerosene lamps to light their premises in the evening, but these lamps are smoky, unhealthy, hard to maintain, expensive, and subject to frequent fluctuations in fuel prices. The kerosene lamps are usually rented out by the food distributor for about $0.35 a day. SELCO offers to replace the kerosene lamp with electric lights and a battery and a solar charger. Because the banks will not lend to the hawkers—they are too poor to offer any bank guarantees or fixed assets as collateral—SELCO works directly with the distributor who sells food to the hawkers on a daily basis. The distributor may now own a solar power plant charging station thanks to financing from a bank and guarantees by third parties. He delivers a charged battery to each vendor with the food and at night collects the drained battery. The lights are fixed to each hawker's cart. The result is that hawkers are now insulated against the rising cost of kerosene and unhealthy kerosene fumes and have reliable lights for $0.25 a day, or less than the cost of kerosene. The food distributor, who collects daily fees for the charged batteries, is able to repay the bank for the loan for the solar charging station within two to three years, thereby making the solar enterprise commercially viable.

THE USE OF SOLAR PHOTOVOLTAICS IN NIGERIA

Despite a large income from oil production and the export of natural gas to neighboring countries in West Africa, Nigeria's electric power network serves only 36 percent of the population, mostly in urban areas and often intermittently. This limited service has a significant impact on nearly all development goals, especially in rural areas. It is more difficult to stem the migration of agricultural producers to the cities. Communications and transport networks cannot be built in areas where cell phones and transmitting stations cannot be powered and traffic lights fail to function. Hospitals and clinics cannot operate without refrigeration and sterile medical equipment. And children are unable to study at home at night.

Because of the great strides made in silicon wafer technology and thus solar photovoltaics over the past decade, the cost per kilowatt-hour has fallen to almost equal the cost of power from the grid. Solar energy can be employed in as small a unit as a house or in as large a unit as a town. It can provide lighting and power small appliances such as phones, TVs, and small refrigerators, and, in larger settings, street lights and traffic lights. Although the price of this technology has been falling as more effective materials are developed, most applications in developing countries, including Nigeria, are subsidized by governments or donors. Commercialization is probably necessary to ensure the widespread and sustainable domestic use of PV technology.

The primary market for solar photovoltaics in Nigeria is rural villages. Another market is the low-income urban neighborhoods that are poorly served by the grid, or where the grid suffers frequent breakdowns. In cities, the hawkers are mostly ambulatory and sell to cars in traffic, and they appear in villages only on market days. Therefore, battery charging may be of only limited use to the hawkers.

In Nigeria, several demonstration projects have been carried out in which solar systems have been installed in rural villages with donor or government support. For example, the University of Nsukka installed a system in a village 7 kilometers from the university. Fifty households, street lights, and a community center were included. The benefits to the village were realized at once, including longer hours of work and higher incomes and more opportunities for recreation. However, because the system was centralized in the village and was free to residents with no apparent owner, within a short time some abuse and overloading occurred, which burned out the system. The university concluded that training of the beneficiaries should be an important component of such projects.

A U.S.-based NGO, the Solar Electric Light Fund (SELF), has been carrying out a demonstration project funded by the U.S. government and the Jigawa state government in northern Nigeria in which several villages are provided with PV units and services. The services include solar-powered water pumps; lighting, cooling, and vaccine refrigeration for the village health clinic; lighting in primary schools for adult education, homework at night, and computers; street lighting; lighting of mosques; and new solar microenterprise centers. However, the U.S. funding has ended, and the project is presently being managed by a local NGO.

Related efforts by private companies are under way. Solar water pumps, inverters, and solar cookers are available locally, and the National Agency for Science and Engineering Infrastructure (NASENI) has a pilot project on the manufacture of solar cells.

HYPOTHETICAL CASE STUDY

To carry out the hypothetical case study, the participants at the Lagos workshop on photovoltaics invented a hypothetical enterprise that would install solar photovoltaics systems in Nigeria, and then it proceeded to create the elements of a business plan.

The enterprise. The hypothetical enterprise is called the Solar Energy Company of Nigeria.

Statement of purpose. To market, sell, install, and maintain solar photovoltaic products for communities not served by the national or municipal power grid, with a special emphasis on rural areas. Initially, the company will emphasize solar photovoltaic home systems.

What is the product or service? A four-point light solar package for home

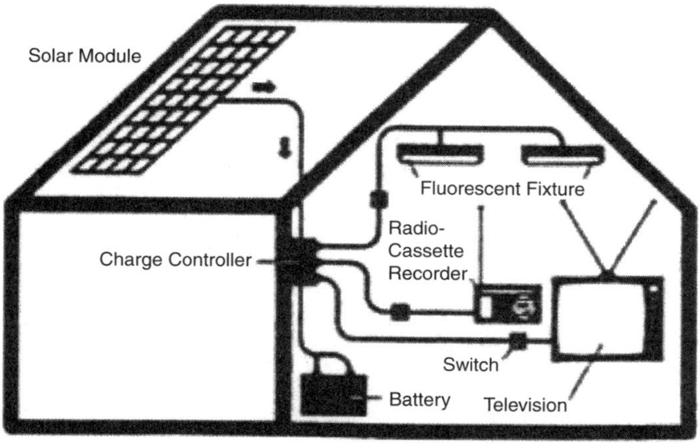

FIGURE A-1 Photovoltaic home system.
Credit: Dr. Harish Hande, Managing Director, SELCO Solar Light(P) Ltd. India.

systems and small enterprises, installed, with a credit package and a maintenance plan (see Figure A-1). The system should be modular so that it can be expanded. In the short term, the wafers and deep-cycle batteries will be imported; 12-volt appliances are available locally. AC power will be available with installation of an inverter, possibly on a separate system so that if the refrigerator fails the lights stay on.

The unit will sell for about 75,000–80,000 naira (about $500–$600) for a 40-watt solar panel, a battery, a LED panel that monitors battery level, and four lights, based on a cost in Nigeria of 40,000 naira for the panel, 10,000 naira for the battery, and 5,000–6,000 naira for the wires and fixtures. This system would be designed to provide four hours of light at night when there is abundant sunshine during the daytime.

Who are the customers? The customers will be homeowners and small enterprises not served by the national electricity grid, especially in rural areas. In Nigeria, this customer base may include as much as 60 percent of the population, or about 100 million people. There may also be niche markets such as vendors who require charged batteries at night.

Who is the competition? Legally, the Power Holding Company of Nigeria (PHCN) is the sole supplier of power to homes and businesses. It is not known whether installers of household solar systems will be challenged by the PHCN.

Shell Oil, which offers sales and installation of solar systems in India, has a presence in Nigeria and may enter the market if it is shown to be profitable in Nigeria. The local medium-size solar companies serving larger enterprises such as hotels could grow to become strong competition in villages in Nigeria.

What technologies and procedures would be used in the manufacturing, assembly, delivery, and service of the product? The four components of home solar installation are the module containing the wafers, the battery, the wiring, and the structure. The wafers can be either amorphous or crystalline silicon, whichever is most cost-effective at the time of purchase. For the short term, these items will be imported, but as local industry develops it will be more economical to buy them locally. The deep-cycle battery, also imported initially, will last up to eight years if well maintained and replenished with distilled water. The manufacturer offers a three-year warranty, but it may be necessary to extend the warranty to the customer and the bank for five years and provide service with distilled water.

Assembly of the unit will be at the customer's site.

A maintenance contract will be required. It can either be included in the price or sold separately. If sold separately, the government may provide vouchers for the maintenance contract, as is done in other countries. Insurance may also be required by the creditor.

An alternative AC system that includes an inverter also can be offered. The system with inverter, usually 50 watts, will cost about 80,000 naira for the solar panel and 26,000 naira for the inverter, acquired locally in Nigeria, or an additional $250.

What are the advantages and challenges for this enterprise in Nigeria? Nigeria has experience and core competency in solar energy. Many small companies are serving niche markets, including solar water heating and drying. Several universities have ongoing research programs, and there have been demonstration projects in villages in several locations. The potential market is huge, arguably up to 100 million people who are underserved by the national grid. Funding may be available for first-stage financing and for credit to buyers from government or even the oil industry. The entrepreneurial spirit is vibrant in Nigeria, and a demonstrated success in one region may encourage imitators in other parts of the country.

Worldwide, a great deal of experience has accumulated on the installation of solar home systems in developing countries. However, it is well known that many of these systems fail, and that the principal reasons are not technical, but failures of the business model and adaptation to local customs and capabilities. These failures include no training of users, no service and maintenance, and no feeling of ownership on the part of the users. The last is characteristic of systems donated by universities or NGOs for demonstration purposes; these systems are typically in operation for less than a year.

A potential local problem in some areas is that users may become discouraged during the rainy season when there is insufficient sun for several months; a complementary program of battery exchange may be a useful paid service during such periods. Another problem may be that customers who are not in the habit of switching off lights or appliances may overuse their systems and run out of power. In the summer or dry season, the dust that may cover the solar panel must be removed daily. It should be obvious that the panel must be kept in the sun, but some users must be trained and reminded to do so. The battery should be kept in a box out of the reach of children; the voltage is low, but the lead and acid in the battery can be dangerous. Distilled water must be added periodically, and this maintenance function should be part of the service contract.

Production requirements. Because most materials will be purchased and because the solar systems will be assembled at the customer's site, it will not be necessary to require extensive space for manufacturing or assembly. The centralized office space needed (about 300 square meters) will be used mainly to store materials; it should be close to the work area with good road access. In urban areas, the cost of such space will be 2,500–3,000 naira per square meter per year; the cost will be lower in rural areas. The office should contain a small, well-furnished showroom and storage space. Solar modules and batteries should be stored in locked cages with camera monitoring for security. Insurance could be expected to cost 2 percent a year, including fire and theft. The materials can be picked up and brought to the workplace by two workers using a motorcycle or small vehicle, which can be rented when required. The length of travel and access to stored materials will determine the time required between ordering and installation of a unit—an important competitive point.

The service package will also be important in both meeting competition and satisfying banks and financiers. Competition and financing will determine whether the service package should be included in the base price or should be an important add-on that provides value-added to the customer at a modest cost. In some countries, the government offers vouchers to homeowners for service contracts as an indirect subsidy to the solar industry to encourage energy savings. In such cases, it is worthwhile to separate the maintenance from the base cost. Alternatively, if subsidies are available for the complete installation package, it would be better to include service in the base price as a necessary component of the package.

Human resources. The enterprise could be started with a minimum of five people: a manager, two engineers, a secretary, and a sales/marketing agent. In a rural area, the minimum salary is about 15,000 naira per month, 25,000 naira in Lagos. An engineer earns about 30,000 naira a month, whether in Lagos or in a rural area.

The marketing agent could be paid 35,000 naira a month, plus commission. Paying a uniformed guard would require an additional 12,000 naira a month. The installations could be carried out by the two engineers equipped with a cell phone who would travel by motorbike. They should be able to install a four-light system in two hours. Thus they should be able to complete two installations and spend two more hours on sales a day, with travel.

Legal and regulatory requirements. Every new company must register in Nigeria, at a cost of about 145,000 naira, including legal fees. Registration, which takes about three months, requires a search of names and registrations and the participation of a lawyer, preferably based in Abuja. An environmental impact statement may be required.

Under Nigerian power company (PHCN) rules, a PHCN subsidiary has exclusive rights to provide electrical service in its area. Thus privately generated electricity may be subject to a legal challenge or a tax. To clarify the point and to encourage alternative energy sources, renewable energy legislation would be valuable. It should allow renewable energy equipment to be imported duty-free, provide businesses with credit for using renewable energy, and arrange for net metering on the electric grid to encourage the private generation of electricity.

Marketing requirements. The market entry point can be urban or rural. Although in the long term rural areas may provide the largest and most stable market, it may be easier to find paying customers and arrange financing in an urban area. Urban dwellers who donate systems to relatives in a rural village may provide an entry point to the rural market. Similarly, installing a system in a chief's house at a reduced cost may influence others in the village to buy. Experience in other countries suggests that thereafter the best marketing will be via word of mouth, and a system installed in a village is the best showroom. In any case, microfinancing will be necessary for rural customers.

Most likely, it will be necessary to approach banks on behalf of clients and to demonstrate the system and describe the guarantees. The Nigerian Association of Small and Medium Enterprises (NASME) in Lagos has a microcredit arm, the Small and Medium Industries Equity Investment Scheme (SMIEIS), that could be helpful. Although customarily it lends to enterprises, the funds might be used by the new company described here to build a rotating fund that finances the purchase of solar systems.

Outreach, education, and training. Training will be an important part of the business—for staff, for customers, and for banks and financial institutions. Training manuals for the sales and service staffs should be designed for classroom instruction, to be followed for service staff by training in carrying out installations in the field with experienced technicians.

For financial institutions, training may be offered on an occasional basis in the form of conferences of bank managers and field officers, led by university professors of engineering hired for the purpose. The banks may develop their own manuals to train, in turn, the inspectors who approve the loans for the installations.

Implementation plan. The steps recommended here reflect the experience of SELCO in India and several Nigerian solar energy entrepreneurs.

Before entering a new area of operations or a new business, an entrepreneur must make certain calculations. Most important are the start-up costs and the anticipated costs of operations, including service, taxes, and travel, depending on the anticipated volume of business. These costs also depend on the target area, the ease of transport, the number and rental costs of office sites, staff salaries, and Internet service (and web site) costs. A would-be entrepreneur also must estimate the cost of inputs, materials, and imported items in order to calculate the price of the product and service contract. That price will determine when the company can expect to begin making a profit. Other decisions that must be made revolve around a strategy for brand building and marketing, the sources and availability of the various components, which products to promote, and how to maintain the quality of the product. The entrepreneur also must prepare a business plan to assist in planning and seeking funding. It may be useful to prepare a five-year plan, and then back it down to the first year. The entrepreneur should anticipate a 10-year period for nucleation and expansion before the market becomes saturated and a new market area or technology is needed for sustainability.

To assist with these calculations, the entrepreneur must estimate the average insolation (hours of sunshine) in the projected target area, which will determine the size of the unit and its potential for satisfying customer need. These data may be available from government, or they could be acquired using simple instruments (although a complete record will require a year). Estimates of the number of families in the area without electric power and the number currently served by the grid but who also might purchase a unit to protect against outages will be needed as well.

It is possible that the government would offer grants or subsidies to a new or established company to provide service to areas not served by the grid. For that reason, it would be worthwhile to prepare a description of the social impacts of the technology to use in requesting support or incentives from the government. Possible incentives include an interest discount or depreciation benefits for the first year (in India this benefit is 80 percent) or a reduction in registration license fees.

More broadly, it may be useful to form a stakeholders' group to advise and assist any individual or company interested in pursuing this initiative. It should include experts from Nigerian universities (particularly the business and energy centers), government, NGOs, and the private sector. The National Center for Women Development in Abuja, the International Center for Business Research, the National Agency for Science and Engineering Infrastructure, and the Nigerian Association of Small and Medium Enterprises would be useful participants. The Nigerian Academy of Science could convene the group, taking care to show objectivity and impartiality toward potential entrepreneurs. The Academy might be well positioned to partner with a university or NGO to offer nationwide courses and other training activities in entrepreneurship and photovoltaic technology and to initiate the presentation to the government in favor of renewable energy legislation.

The many international NGOs with broad experience in solar photovoltaics in rural areas also could be called on for advice and assistance. Among them are Solar Energy International (SEI), SunEnergy Power Corporation (SEPC), Green Empowerment, and the Solar Electric Light Fund, all in the United States.

Financing. For most entrepreneurs unaffiliated with larger enterprises, first-stage financing will be required. Several sources in Nigeria might have a particular interest in businesses of this sort. The just-mentioned Small and Medium Enterprises Equity Investment Scheme of the Nigerian Association of Small and Medium Enterprises offers equity loans. The association takes an equity position in the borrowing firm, which it can convert or sell after a fixed time period. Twenty billion naira are available per year for bankable projects. The association also may provide entrepreneurs with references to commercial banks. Some insurers will require an insurance policy, which must be included in the cost of operation.

The other critical type of financing is a credit system for purchasers. Banks, both formal and informal (market), are often seeking to make loans, and, with some training, they may see the benefits of microloans, secured with operational solar systems. Currently, most loans are for one year, and the interest rate is not a critical parameter. (In India, SELCO began to work with interest rates of 24 percent.) Regional banks, cooperative banks, and other forms of cooperative investment also may be worth approaching. Alternatively, a cooperative society or NGO working in the villages may buy in quantity for its members or clients. However, this relationship must be considered with care, because experience has shown that if the solar unit is donated to the household, there is a high probability that it will not be operated or maintained properly.

Partnerships. Partnerships may be important for the success of the new enterprise. Partnerships are helpful for marketing, finance, and possibly purchasing or manufacturing materials and technologies. Potential partners include NGOs, state governments, cooperative societies, financial institutions, and other businesses.

In the short term, solar modules and deep-cycle batteries will have to be imported to Nigeria, but in the longer term they could perhaps be manufactured domestically. Nigerian scientists have been active in initiatives with foreign partners that make use of amorphous silicon and, in the longer term, might use organic materials, so-called polymer electronics. The government has promised support. A joint venture with a local manufacturer of solar cells and batteries could have positive benefits for both parties.

PARTICIPANTS, SOLAR ENERGY WORKSHOP

Ike Abugu	Nigerian Association of Small and Medium Enterprises
G. Adetunji	Obafemi Awolowo University
O. Adewoye	National Agency for Science and Engineering Infrastructure
Alex Animalu	University of Nigeria, Nsukka
E. Ene-Ogong	Cross Rivers State University of Technology
Njidda Gadzama	Nigerian Academy of Science, University of Maiduguri
Michael Greene	U.S. National Academies
G. B. Ogunmola	Nigerian Academy of Science, University of Ibadan
V. O. S. Olunloyo	University of Lagos
O. Oparaku	University of Nigeria
M. R. Pai	Solar Electric Light Company (SELCO)
Walt Ratterman	SunEnergy Power Corporation
A. Salau	Obafemi Awolowo University
R. I. Salawu	University of Lagos
Wole Soboyejo	Princeton University
B. Sogunro	Jombas Ventures
H. Udemba	Prime Energy Ventures
Joshua Uzoukwu	Juneko Engineering

Appendix B

Water Purification: Hypothetical Case Study

This hypothetical case study on water purification explored the opportunities and challenges inherent in investment in a new enterprise that would capitalize on two distinct technologies: one to be sold for use by the family in the home and the other to be franchised or sold to water providers at the village or community level. At present, no company in Nigeria is engaged in these activities, although some companies are selling bottled spring water in the Nigerian market.

Workshop participants included representatives of the Federal Ministry of Agriculture and Water Resources, the Federal Institute of Industrial Research (FIIRO), the University of Lagos, and the University of Ibadan. The U.S. National Academies were represented by Ron Rivera, international coordinator, Potters for Peace of Managua, Nicaragua; Tralance Addy, president and CEO, WaterHealth International of California, United States; Wole Soboyejo, Princeton University and a member of the U.S. National Research Council committee for the project; and Michael Greene, project staff director at the National Research Council. The Nigerian Academy of Science was represented by Gabriel Ogunmola, president, and Njidda Gadzama, vice president. A complete list of participants appears at the end of this appendix.

TECHNOLOGY OF WATER PURIFICATION

Many effective technologies are available for home or community-scale water purification. Some are chemical-based, such as treatment with chlorine. Some are thermal, such as boiling and solar heating. Others are mechanical, such as ceramic filtration. And some utilize radiation, such as ultraviolet (UV) treatment. Criteria for selection of the technology for the hypothetical case study included low cost, ease of assembly or manufacture, safety and simplicity in home operation, maintainability and sustainability, and worldwide experience with successful commercial exploitation. Some of the thermal methods are nearly free (the SODIS system requires only a blackened soft drink bottle and four hours on a sunny roof), although they may require extensive training so that users acquire the discipline necessary for health protection. Because there is no clearly preferred technology for efficiency and commercial opportunity, two distinct business models were studied. It is hoped, however, that a potential investor or entrepreneur will realize that the business models can be applied to a variety of technologies.

The two technological models described here are ultraviolet disinfection, as utilized in the business model of WaterHealth International (WHI) of California for sales in developing countries, and ceramic filtration, as promoted by Potters for Peace of Managua, Nicaragua. The business models are quite different.

The patented UV Waterworks (UVW) ultraviolet disinfection unit is at the core of the large systems sold by WaterHealth International to franchisees who use them to produce and sell potable water in containers to consumers at prices below those of bottled spring water. Other WHI products, such as community-scale water systems, are sold to governments or communities directly, and provide enough safe water to meet nearly all daily domestic needs, including hand and food washing and bathing. The operating costs for a system that can serve at least 3,000 people are less than $4 per person per year. Treated water is sold to recover the investment and maintenance costs of these systems at prices that are within the reach of the populations being served.

The Filtron ceramic filter technology is promoted by Potters for Peace and marketed by individual filter producers directly to households for operation by the end user. It provides safe drinking water for a family of six at a cost of about $7 for two years.

Ultraviolet Filtration

Figure B-1 is a schematic diagram of the UVW unit, which was invented by Ashok Gadgil of Lawrence Berkeley National Laboratory in California and licensed to WHI.

In the UVW system, the UV source is suspended above the water being treated rather than submerged in the water. Water passing through the system is irradiated at high intensity amplified by reflection. This

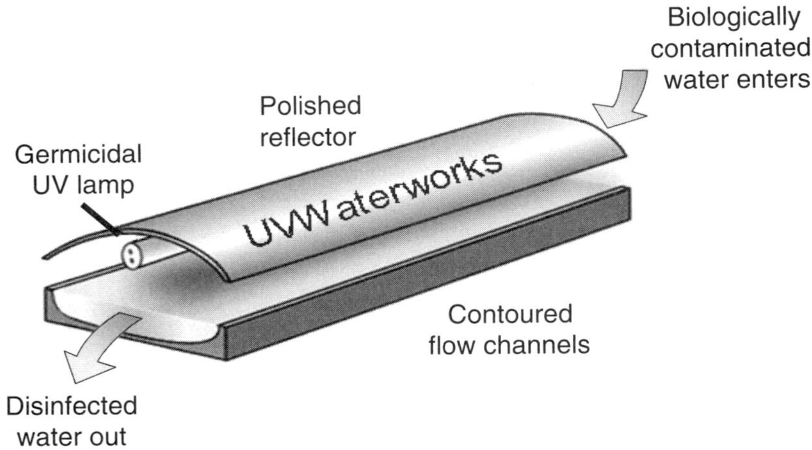

FIGURE B-1 UV Waterworks water filtration unit.
SOURCE: WaterHealth International, http://www.waterhealth.com.

configuration improves performance in challenging environments and overcomes operating barriers, such as corrosion, associated with conventional UV treatment technologies. The lower maintenance requirements enable the UVW system to be operated in areas where labor pools may lack technical knowledge or specialized education.

The community water system can provide safe potable water for 3,000 people. WHI claims third-party laboratories have validated that the UVW technology, operating at a flow rate of four gallons per minute, eliminates at least 99.9999 percent of the bacteria and viruses that cause water-borne disease. WHI applies the technology to a wide spectrum of uses, from homes, schools, and hospitals in developing countries to residences that rely on well water. Very little maintenance is required—the lamps must be changed once a year, and the filters must be periodically backwashed and replaced. UVW technology is also designed to be fail-safe. If any type of malfunction occurs, such as a power outage or a drop in radiation dosage, an automatic valve closes the entry port to the device, ensuring that contaminated water cannot flow through the system without being disinfected. WHI's systems can be powered by solar or wind energy or a generator. In an emergency, they can be run for two weeks at a time with a car battery.

Ceramic Filtration

Figure B-2 is a schematic drawing of the colloidal silver-enhanced earthenware ceramic water filter first developed in 1981 by Fernando Mazariegos of the Central American Institute for Research and Industrial Technology (ICAITI) in Guatemala. The goal was to develop a low-cost filter, producible at the community level, that would both clarify turbid water and make the water safe from bacterial contamination. The technology is now promoted worldwide by Potters for Peace, which offers technical assistance to small companies that wish to produce and market these filters. There is no license fee.

The filter is primarily intended for household use, ideally as part of a water delivery network that provides intensive educational efforts aimed at improving water hygiene overall. The most economical model consists of a porous clay filter unit perched inside a lidded 5-gallon, spigoted receptacle of plastic or clay.

FIGURE B-2 Colloidal silver-enhanced earthenware ceramic water filter.
SOURCE: Drawing provided by SANITEC, Cuba.

The clay for the filter is mixed with an equal volume of sieved sawdust, rice husks, or peanut husks, mixed with water, and then pressed into an aluminum mold with a truck jack and fired in a kiln. The sawdust burns out, leaving 0.06–0.2 micron pores that will eliminate all bacteria and parasites. The filter is then painted inside and outside with colloidal silver, similar to that used in the newer refrigerators, which oxidizes the bacteria and inhibits regrowth, thereby acting as a germicide. The unit must be brushed periodically on a clean surface to remove particles of turbidity. Users of this equipment must be trained in hygienic practices to avoid recontamination of the water.

The unit has a flow rate of about 2 liters of water per hour, which will provide drinking water for a family of five to six persons. Another model processes 6 liters per hour. Potters for Peace claims the filter has been successfully laboratory tested in over 10 countries on four continents, and it has been proven effective in eliminating coliform bacteria, parasites, amoebae, and *Vibrio cholera* from water.

APPLICATION EXPERIENCE

Ultraviolet Filtration

WaterHealth's systems are distributed using different business models in different countries, but always by means of local sales agents. One of WHI's earliest commercial successes was in the Philippines. In 1997 WHI established a subsidiary in Manila to offer franchises that would use the UVW technology to provide lower-cost alternatives to bottled water. The subsidiary enables local "mom and pop" store proprietors to own and operate WaterHealth-branded "water stores" that would benefit from WaterHealth's expertise—such as on where to open their stores, how much foot traffic versus deliveries they should expect, and how to manage finances—as well as technical services. Water is sold from storefronts in sanitary containers.

Today, over 3,000 water stores are vying for business in Manila's thriving market. The WHI franchisee pays about $8,000 for a turnkey operation plus franchise fees. All franchisees are trained to operate their stores in compliance with the highest sanitary and quality standards. The Philippine government regulates water stores, and in order to maintain a "Grade A" facility, as required by WHI, a WHI water store must occupy at least 20–30 square meters of storefront. Access to municipal or well water and electricity is also necessary.

The franchisees recover their costs after a period ranging from 10 months to two years. The water store franchise model establishes a commercial model and entrepreneurial opportunities for those who would like to own and operate their own businesses, while at the same time giving something back to their communities. Today, roughly 50 WHI franchisee water stores are in operation in the Manila area, in addition to several water co-ops that were installed with WHI technology via the Rotary Club there.

In other countries, WHI is establishing an alternative model, the community water system (CWS), to serve small villages in areas once thought to be unreachable by piped water. Figure B-3 shows a rendering of the prefab CWS structure soon to be built in India, as well as a diagram of the configuration of equipment inside.

The CWS shown in Figure B-3 is designed to provide a community of up to 3,000 people with up to 20 liters of safe water per person per day. Systems are modular and scalable—that is, they can be configured easily to serve larger or smaller populations.

In India and Mexico, WaterHealth markets the CWS as a "micro-utility" to governments and village organizations, which recover the investment in the community water systems through the sale of treated water to villagers. For the purchaser, it is a turnkey operation, including treated water storage and the pipes to bring the water from up to 2 kilometers away to the village for treatment if necessary. WHI hopes to integrate rainwater-harvesting capabilities into future installation designs.

In India, WHI created WaterHealth India, which seeks local affiliates in each state. In Andhra Pradesh, WHI sells equipment to the Naandi Foundation, which markets the units, with a service contract, to its own local franchises, entrepreneurs, and village organizations. The village organizations typically sell coupons to families to redeem for water; end users may opt to pay extra for home delivery in special containers. Part of each community water system is an educational program, usually conducted by a nongovernmental organization (NGO) under contract, that encompasses health and hygiene issues and encourages people to use clean water.

The cost of this turnkey program is about $50,000, or about $17 per person, which is lower than alternatives of similar capacity, such as bore wells that normally provide no disinfection, filtration, safe storage, or education on health and hygiene. By comparison, municipal facilities cost around $100–$250 per person to build, and the costs of bore wells are similar. After a community fully adopts a CWS, it should be able to generate $24,000 a year acting as a small utility.

WaterHealth International, as a U.S. firm, is sometimes able to take advantage of the U.S. Export-Import Bank to loan money to the local subsidiary to buy U.S.-made equipment. In India, WHI has arranged with an Indian bank for the loans to finance the installation of a large number of community water systems.

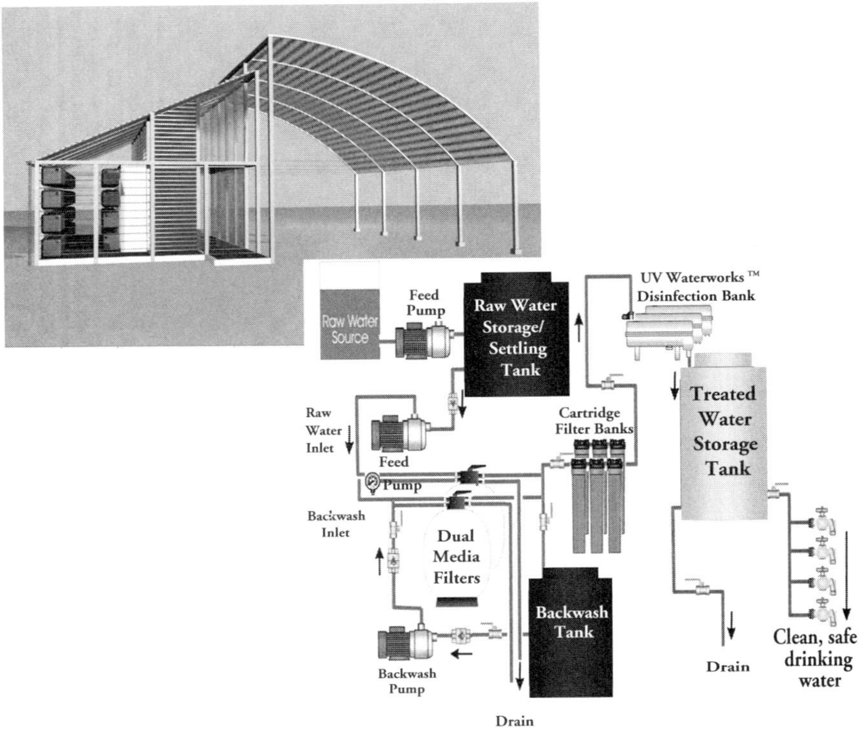

FIGURE B-3 Prefabricated community water system structure to be built in India. SOURCE: WaterHealth International, http://www.waterhealth.com.

Ceramic Filtration

Potters for Peace (PFP) was formed in 1986. In October 1998, when Central America was struck by Hurricane Mitch and, as part of the relief effort, the Red Cross was importing clean water in large quantities, Potters for Peace sought a way to produce the water locally, using low-cost existing technologies. It chose a model that had been developed at ICAITI in Guatemala in 1981. As noted earlier, it sells for about $7, and is usually replaced after two years. It removes 98 percent of impurities and infectious organisms. Potters trained by PFP began by selling the filters to NGOs, which then distributed them in the villages. But this model was not sustainable because the NGOs tended to donate the filters to households, which thus were not motivated or trained to use and maintain them. Meanwhile, the donations undercut the market for commercial sales.

Presently, the filter companies set up by PFP produce about 1,000 filters per month, sold mainly to NGOs in Nicaragua. PFP charges the local organizations for training on a sliding scale, depending on economic and social criteria. PFP's income is supplemented by consulting fees from donors and clients as it helps small companies in other countries produce and sell water filters. The cost of the product around the world depends mainly on the cost of labor plus fuel or electricity, and the price varies between $5 (in Nepal) and $25, or even $100, for designer models that have the same water-purifying effectiveness but feature elegant containers designed for a more affluent clientele.

The cost of a factory that employs two to four people is generally about $10,000, and this cost can be recovered in as little as nine months. Often the producer partners with an NGO and receives the benefit of financial and marketing assistance. Potters for Peace offers the plans of the press and other equipment at no charge, or it will refer the client to mechanics who will build a press to order at a cost of $800 for a portable system able to produce about 50 clay filtering elements a day. Today, 100,000 Filtrons are in use throughout the world, serving about 500,000 people.

WATER PURIFICATION IN NIGERIA

In most countries, municipal water is the responsibility of the government and is subsidized for the consumer. The maintenance of pipes, pumps, and other infrastructure is generally a problem everywhere. In Nigeria, the problem is particularly acute, and it is widely believed that no community, rich or poor, in Nigeria receives safe, drinkable water from the government. Even those who rely on deep bore wells cannot rely on safe water, and so bottled spring water is widely used. The World Bank has sponsored several water projects, but most of these have been urban; rural areas are overlooked.

As a result of this situation, some Nigerians have never had clean water, and others drink only bottled water. Water is sold on the street in containers ranging from sachets in plastic sacks to bottles of mineral water. Many social programs exist to provide water, but the social agenda has not been linked with a business model to provide sustainability.

Nigeria has 320 million cubic meters of water of which 86 percent is surface water. Forty percent of the population, mostly in rural areas, has access only to surface water. The water table ranges from 300 meters in sedimentary areas to 70 meters in basement areas. Water can be found above 70 meters, but the salinity levels are high.

The Federal Ministry of Agriculture and Water Resources is responsible for Nigeria's water policy. The ministry has links with the United Nations Development Programme and the International Atomic Energy Agency for resource surveys and isotopic analysis. Amadeo Bello University has a mass spectrometer for contamination studies, and reference labs can be found in Lagos, Enugu, and Taiguna. The ministry provides assistance digging deep bore holes, and it provides hand pumps, bicycle-powered pumps, rain-harvesting technologies, and in-service training on maintenance and use. It advises consumers to use water from deep wells, with charcoal and sand filters, for domestic water. The government requires a minimum distance between wells and septic tanks, but the requirement is hard to enforce. Shallow wells are often contaminated by informal backyard cemeteries.

Water sellers often get their water from wells. Some is boiled or treated with ultraviolet radiation. Some facilities are inspected, but the water is not tested. Commercial bottlers sometimes get water from industrial steam as well as from natural springs. But even bottled water may be contaminated by users after purchase, for example, by adding locally made ice. Street water sells for 5 naira for a 500-milliliter bag and 40–60 naira for a bottle. In Nigeria, safe water is more expensive than petrol.

It is said that no one in Nigeria drinks piped water. In many rural populations, there has never been an alternative to drinking contaminated and sometimes turbid water. Some potential users argue that the resistance to waterborne diseases people have developed will be lost if clean water is provided. But children under five have no resistance, and many people suffer from diarrhea as a result of drinking impure water.

Education will play a critical role in the adoption of the technology and safe water-handling practices in the communities served. When people understand the high costs associated with drinking unhealthy water, including ill health or death (especially of infants and young children), loss of working hours, and hospital fees, they will realize the economic benefit of paying a small fee for treated water. The fees can support profitable companies and make the system sustainable.

The government of Nigeria has embarked on a costly program of urban water purification using mass spectrometry for separation as well as filtration and chlorination. It plans to repair the existing corroded pipes. Impurities in water include chemical and biological contaminants and volatile organics from plastic pipes, which are thought to cause infertility and a decline in the birthrate. Commonly diagnosed waterborne diseases include diarrheal disease, dracunculiasis (guinea worm), onchocerciases (river blindness), and typhoid.

HYPOTHETICAL CASE STUDY

Participants in the Lagos workshop on water purification created a hypothetical enterprise to install water purification systems in Nigeria and the elements of a business plan.

The enterprise. Because the workshop considered two very dissimilar technologies, participants had to decide whether to imagine two distinct hypothetical companies or one. Filtron production is relatively quick to organize at low cost; a WaterHealth-type operation would take longer and would be costlier to assemble. The workshop decided to discuss them together as one company, recognizing that an investor can pick and choose the elements that suit real opportunities. The division that makes and markets ceramic filters might be called Filtron International.

Statement of purpose. Provide potable water and associated services affordable to everyone, specifically including those in rural areas and the urban poor. The water would be provided by direct sales, through ultraviolet filtration, or indirectly, through sales and service of home ceramic filters.

What is the product or service? For the Filtron, the product is a ceramic filter in a suitable containment vessel. For a small additional cost, a tester like those produced by the Hach Company of Colorado could be included to test for bacteria and other contaminants to assure the user that the unit is functioning properly.

For the UV procedure, the product is both equipment and potable water. Equipment is sold to franchisees, independent agents, local communities, or governments that produce safe water to be sold to the public. A maintenance contract on the unit is offered to the purchaser, and educational programs for the consumer may be offered to the communities. Treated water is marketed and sold in safe containers, perhaps bearing the company logo.

Who are the customers? The ultimate beneficiaries are any people who lack access to potable water, specifically including those in rural areas for whom municipal water service is not available and the urban and periurban poor who cannot afford well water or bottled water. Potable water or Filtron units also can be offered to middle-class customers in more elegant containers, and bottled water can be marketed through water stores or for home delivery. Branded water also may be sold in bottles in city traffic, with some security measures applied to discourage competitors with counterfeit (nonpurified) brands or contaminated containers.

In rural areas, operators of community water systems or water refilling stations may set up facilities beside a river or another traditional source so that families would have the option of buying safe water in portable, reusable containers. By selling other products and services, ranging from laundry soap to Internet access, entrepreneurs could turn the site into a small commercial hub.

The Filtron ceramic filter has been profitably marketed to NGOs and donor agencies in Central America, Cambodia, and other places for distribution in needy areas. Such marketing must be done with care, however, because NGOs often donate the filters to households at no cost and with minimum training and service, and they frequently fail for lack of maintenance. Such donations could seriously undercut the direct sales market.

What materials, technologies, and facilities would be used in the production, delivery, and service of the product? Filtron ceramic filters and UV radiation systems are manufactured in different facilities and can be discussed separately.

Ultraviolet disinfection. UVW-type units would initially be manufactured elsewhere and imported to Nigeria. It is possible that, after the development of a sufficient order pipeline, the units could be manufactured in Nigeria. The tank, containers, flow meter, pumps, PVC plastic casing, and spare parts can be made or sourced locally. Cone and filter paper cartridge filters can be acquired locally. Test equipment or laboratory services also can be arranged locally. At least 1.5 kilowatts of power is needed to operate a community water system. Depending on the water source, lower power requirements may apply to water store franchises that do not have to pipe water to their facilities. Solar cells and batteries for this purpose can be imported if desired, and in the future may be available locally.

Ceramic filters. The skills, tools, and materials needed to manufacture the Filtron filter are common to ceramics workshops everywhere. Clay is combined with sawdust, peanut shells, millet straw, or similar material in equal parts, resulting in 25 kilograms of clay to 6 kilograms sawdust. The mix is pressed into a mold, dried, and then fired in a kiln to 890 C. After tests on initial filtration rates, colloidal silver is painted on.

During the firing process, the sawdust burns off, leaving small twisted pores through the ceramic material too small to permit bacteria to pass. Colloidal silver provides additional antibacterial properties and prevents regrowth. The colloidal silver can be imported from Spain, Germany, or Mexico at a cost of $60 per liter, enough for 500 filters. It might be possible to manufacture colloidal silver in Nigeria, but the manufacture of Filtrons, at about $0.12 worth of colloidal silver per filter, would hardly support a local industry alone.

Quality control is important, and the filtration rate and bacterial count must be monitored. The latter can be performed with the Hach presence/absence test, using sachets that change color to reflect the presence of pathogens. The cost is about $0.50 per use.

The ceramic filters could be made from locally available clays, wood chips from local saw mills, and rice or peanut husks from local villages. The containers for the water also could be made locally, using local facilities for pottery and plastic manufacturing. Thus the capabilities exist within Nigeria to manufacture complete Filtron ceramic filters from locally available materials. This was demonstrated at the end of the workshop when Ron Rivera of Potters for Peace collaborated with Ishmael Bashir, a ceramist at the Lagos-based Federal Institute of Industrial Research (FIIRO), in making filters with an aluminum mold, using a kiln and ceramic processing materials that were available in Bashir's lab. The real challenge is to combine this technical expertise with effective management that can manage and scale up the production, delivery, and service of the filters.

What are the advantages and challenges for this enterprise in Nigeria? The factors in favor of success are the large potential market, the 100 million people who do not have access to potable water, and the culture of business and entrepreneurship in Nigeria, which may lead many people to attempt to enter this business, including many water sellers who presently lack the technology to purify the water they sell. In the case of Filtron, all material and labor, except the colloidal silver, can be obtained locally.

The greatest challenges stem from the same factors as the advantages. The large potential market includes many people who have a history of consuming contaminated and turbid water and so do not see it as a cause of their health and infant mortality problems. In some areas, people may refuse to pay for water because they consider it a natural right, just as some Nigerians view the oil produced in their country. An educational campaign is needed to activate the market. The large number of potential entrepreneurs practically ensures that within a short time after a business appears to be successful, or even before, copies or similar products will appear on the market. The same technology should not be provided to potential competitors in the same area. To protect the market for Filtron, use of Hach tests should be encouraged by including test kits with the filter. Instructions for use of a Filtron should be displayed on the unit with waterproof paper or imprinted directly on the container to discourage imitators.

For UV-treated water, the difficulty will be greater, because the product will come into direct conflict with traditional water sellers, most of whom do not take care to purify adequately the water they sell—and they may be able to sell at a lower price. Some may begin to mimic the brand. Steps must be taken to brand the product clearly in a way that is hard to imitate, and to emphasize safety and reliability in all marketing activities.

The greatest challenge may be getting first-stage financing, especially for the UV system. The Filtron requires the facilities and skills of a ceramics company, and the candidates likely to enter this business are those already engaged in ceramics manufacture. The UV system enterprise will be engaged in selling and delivering water, and may find support from the government or from commercial banks.

Production Requirements

Ultraviolet filtration. The units can be assembled in a central location within the region. A thorough market assessment would be needed to predict the minimum order sizes in order to determine the size of the local business. Local agents or distributors would be identified in each region or state. Field sites or franchises should be selected for marketing advantages and access to a water source, whether surface water such as a river bank where people are accustomed to coming for water or ground water in a site where a bore well may be feasible.

The franchisee will have to acquire or build an attractive facility, stable and rainproof, near the water source or easy for customers to access. It must be about 20–30 square meters in size to accommodate production and the storage of clean water. Tanks for the storage of river or well water may also be required, along with at least 1.5 kilowatts in electrical power from the grid or solar power. A staff of two or three can operate the water store and serve customers; larger delivery areas may require additional delivery staff.

Maintenance service is perhaps more important than the choice of technology. In fact, the project can be considered turnkey only to the extent it includes maintenance and service. The service organization must be based locally; for example, WHI will not enter a market without first establishing the ability to provide adequate service. The franchisee must stock the essential spare parts, including UV disinfection units and other parts not available in the country. Once a month the unit must be opened and cleaned with a wet rag, water samples sent to the lab, and a new certificate issued, which should be prominently displayed. The UV bulb must be changed once a year. Experience has shown that poor people will pay for safe water, but they expect a demonstration that it is clean.

Ceramic filters. Production of Filtron units requires a well-equipped ceramics facility, with a kiln of capacity 1.5 cubic meters. A 16–20 ton press, or alternatively a truck jack, a hammer mill, and a pug mill to extrude the clay, are needed as well. Plastic receptacles and faucets should be ordered locally.

Human Resources

Ultraviolet treatment. The company must conduct a thorough market assessment before mapping out staffing operations. Staff would likely include a chief executive officer, an operations director, a quality control officer, and technical staff to assemble the UV disinfection units. A staff person would be placed in charge of marketing, urban and rural, and this person would require transport. The maintenance service for the equipment could be contracted out to an experienced service company, but training would have to be provided. The quality control officer would work directly with the vendors. It is advisable to fill these positions through advertising and application.

For the water store franchise model, the franchisees could be recruited through advertisements and trade shows. Franchisees would sign a contract with the company that provides for purchase or lease of the equipment plus a franchise fee and monthly royalties. The franchise agreement should specify all details, including hand washing, dress code, hygiene, equipment maintenance, and water testing, if the latter is not included in a service contract. The quality control officer would independently test the water and inspect the equipment. A typical water store franchise would likely require three employees for filtering water, washing and refilling containers, and water delivery, depending on the area served and its demographics.

Ceramic filters. The ceramic filter factory would require at least two experienced potters, one skilled in kiln use and temperature control. Staff should be devoted to marketing and distributing the filters, and others to preparing written and oral materials for instruction for users. This business is usually grafted onto a working pottery facility, which should be prepared for a major expansion of sales and distribution of its low-price, life-saving product.

Legal and regulatory requirements. Presently, all surface and ground water is the property of the state. The authority to distribute water is exclusive to the state government, and it may be illegal to bring water into a home without permission and without paying certain taxes and fees. Normally, the government does not implement the rule, but one homeowner in Ibadan, who is connected to the municipal water supply but has never received any water from its pipes, was threatened with prosecution for bringing water into his own home without permission. It should be clarified whether there is a right to sell water to homes. Currently, a bill is before the parliament to clarify a similar situation related to electricity.

Insurance requirements are also unclear, and may depend on the provisions of the financier.

An environmental impact assessment would have to be prepared. The use of UV and colloidal silver must be approved by the Ministry of Health, and a certificate for the use of UV will be required. The water product must be approved by the National Agency for Food, Drug Administration and Control (NAFDAC), and a NAFDAC number will be required, at a cost of about $1,000.

Marketing requirements. Marketing would likely take place at two different levels. The first would be general to make a compelling case, especially to poor people, for the purchase of water or a ceramic filter in order to upgrade their quality of life and avoid the negative consequences of poor health. The second would be to brand and identify the product. The role of science must be emphasized in order to distinguish the product from unapproved and inferior competitive products. (In this context, it is important to filter out turbidity in the water, as both models do, to distinguish the product from boiled and chlorinated water and so that people can see directly the effectiveness of the filtering process.) The government and NGOs should be enlisted for the first campaign. An endorsement from the Ministry of Health would be effective, if it can be persuaded to admit that municipal water is not adequate. The government should be urged to subsidize diarrhea prevention as it does HIV prevention. "Wash your hands, filter your water" would be a good slogan for a government campaign. NGOs could participate through the Internet and at health centers, especially maternity and pediatric wards, where NGO employees may be working. Radio might also be effective.

As noted earlier, people who have never had potable water may not want to pay, because either they do not see the need or they believe it is a natural right. It would help to ensure that the vendors are local people, and that the emphasis is on the product as not simply water but the process of purification. The low cost of treated water versus the avoidably high costs of consuming untreated water should be analyzed, and the results conveyed through social marketing and education campaigns.

The first marketing decision to be made is the market entry point. Should marketing begin with the highest-cost or lowest-cost product—

that is, with the urban water stores or the rural vendors, the middle-class designer model or the basic Filtron? Coca-Cola is famous for marketing first to the affluent to exploit aspirations. People of means will not be attracted by what appears to be a poor people's product, even if the quality is good, so designer models would be suitable for that market. Meanwhile, the basic model can be sold to NGOs that serve the poor and might be enlisted in the educational campaign. But reliance should not be placed entirely on NGOs, because people may not continue to buy when the NGO leaves the area. Similarly, it is important to prevent the product being given away at no charge except in an emergency. It may be possible to offer basic Filtrons at 1,000 naira ($7) and designer models at 3,000 naira ($20).

The extent of marketing would be limited or expansive, depending on the targets and the development goals of those entering this business. If the goal is to sell a ceramic filter to every home in Nigeria or to dominate the market for bottled water in Lagos in 10 years, marketing would be organized differently than if the goal is to sell 1,000 Filtrons or open 50 water stores a year. Normally, a marketing agent receives a salary of about 35,000 naira, or about twice the average wage of a full-time employee. In Nigerian pharmaceutical firms with global ambitions, the successful marketing manager receives from 2 to 3 million naira a year plus car, health insurance, and housing allowance. In marketing, it is believed, you get what you pay for.

Financing. Both projects of this hypothetical company have been deemed "bankable" or worthy of financing by the chief executive of a bankers' association. External banks, such as the U.S. Export-Import Bank, may provide credit, but they may require overseas purchasing, which is not appropriate for the Filtron component. Local banks would charge a higher interest rate, about 25 percent, and would require capital investment by the borrower. The amount of capitalization required by a bank would depend on the risk and time required to recover costs. That amount may be as high as 80 percent because little of the investment is for capital assets, but mainly to build the company. One possible strategy might be to raise money from the backers of the company to build an initial supply of Filtrons or UV units for demonstration and to meet initial demand, and then go to a bank with the business plan.

Partnerships. For the Filtron effort, a natural partner is the Federal Institute of Industrial Research, which operates an industrial ceramics laboratory capable of turning out about 40 filtering elements a day. FIIRO, in turn, could offer technical assistance to small ceramics companies interested in producing the filters.

The Ministry of Health could be a very effective partner for both marketing and helping to resolve the legal dilemma about the rights to sell water. The Ministry of Agriculture and Water Resources, which is directly responsible for water policy, should also be approached as a potential partner. Among NGOs, the Nigerian branch of the Red Cross may be persuaded to purchase Filtron units to be available in an emergency.

Implementation plan. In view of the importance of safe potable water to public health in Nigeria, it may be useful to form a stakeholders group to advise and assist any individual or company interested in pursuing this initiative. Useful members might include experts from academia, government, NGOs, and the private sector, including the Ministry of Health, the Federal Institute of Industrial Research, the International Center for Business Research, the National Agency for Science and Engineering Infrastructure, and the Nigerian Association of Small and Medium Enterprises. The Nigerian Academy of Science could serve as convener of the group, taking care to maintain objectivity and impartiality among the interested companies. The Academy might be well positioned to partner with a university or NGO to offer nationwide courses and other training activities in entrepreneurship and water purification technology. It could also urge the government to clarify the law about the right to provide potable water to households and to mount a campaign in favor of filtered water to combat diarrheal disease.

The Filtron could be demonstrated publicly at an early date. The ceramics laboratory of the Federal Institute of Industrial Research has the capability to produce a prototype Filtron within a day (provided the colloidal silver is available). If FIIRO finds it difficult to obtain colloidal silver, it might arrange with the suppliers to represent the manufacturer in Nigeria.

PARTICIPANTS, WATER PURIFICATION WORKSHOP

Tralance Addy	WaterHealth International
T. A. I. Akeju	University of Lagos
Ishmael Bashir	Federal Institute of Industrial Research FIIRO
Fola Esan	University of Ibadan
Njidda Gadzama	Nigerian Academy of Science, University of Maiduguri
Michael Greene	U.S. National Academies
A. Adio Moses	Nigerian Academy of Science
O. O. Ogbede	Federal Ministry of Agriculture and Water Resources, Abuja, Nigeria
G. B. Ogunmola	Nigerian Academy of Science, University of Ibadan

V. O. S. Olunloyo	University of Lagos
M. R. Pai	SELCO Solar Light Co. Ltd.
Ron Rivera	Potters for Peace
Folasade Segun	Neimeth International Pharmaceuticals Plc.
Wole Soboyejo	Princeton University

Appendix C

Artemisinin-Based Malaria Therapy: Hypothetical Case Study

This hypothetical case study on artemisinin combination therapy (ACT) for malaria is intended to explore the opportunities and challenges inherent in investment in a new enterprise that would capitalize on a potentially large market for ACTs by taking one or more of the following steps: growing the source plant, *Artemisia annua*; extracting the active ingredient; manufacturing the drug using the highest-quality processes; coformulating the artemisinin component with the partner antimalarial; and marketing in Nigeria. At present, no company is engaged in all these activities, although there are companies, some of which participated in the meeting, that do produce other antimalarials and have some experience with artemisinins.

Participants in the workshop included persons associated with the University of Ibadan, University of Calabar, XeChem Inc., Neimeth International Pharmaceuticals, the Drugs for Neglected Diseases Initiative, Médicins sans Frontières, and Office Technique d'Études et de Coopération Internationales, Paris (OTECI). The U.S. National Academies were represented by Nancy Bearg, a member of the U.S. National Research Council committee for the project and chair of the workshop; Hellen Gelband, Institute of Medicine; and Michael Greene, project staff director at the National Research Council. The Nigerian Academy of Science was represented by Gabriel Ogunmola, president. A complete list of participants is appears at the end of this appendix.

In its use of the knowledge assessment methodology, this hypothetical case study was unusual in many ways. The methodology is most often used to explore the advantages and challenges of creating enterprises that do not currently exist in a country. In this case, however, even though *A. annua* is not yet commercially grown in Nigeria and ACTs are not widely produced and marketed there, companies are actively striving to do so, including some that participated in the meeting. Moreover, the future market for ACTs in Nigeria may be heavily influenced by external factors, such as broad subsidies by bilateral and other donor agencies or the production of a synthetic equivalent of the active ingredient. Without a major global subsidy, the ACTs market is likely to grow slowly and uncertainly, whereas with a subsidy the market could be transformed. Unfortunately, whether current planning will result in such a subsidy cannot yet be predicted. This uncertainty is, in fact, part of the landscape that must be negotiated in the real world.

THE USE OF ARTEMISININ DERIVATIVES

Artemisinin, the active ingredient extracted from the leaves of the *Artemisia annua* plant, has been used for centuries in Chinese traditional medicine to treat fever. Chemical derivatives of artemisinin, in combination with other antimalarial drugs, constitute the artemisinin-based combination therapy, or ACT, which is the most effective treatment for falciparum malaria, the most dangerous form of the disease caused by the *Plasmodium falciparum* parasite.

Malaria kills more than a million people a year, mainly children. Since the seventeenth century, malaria has been treatable, but, historically, there has never been an adequate supply of effective drugs for more than a small percentage of the sufferers. In the 1960s, chloroquine, a synthetic drug, became available at low prices in tropical countries and was effective for decades. Eventually, however, parasite resistance to chloroquine evolved in Asia, and it has now spread throughout Asia and Africa. Chloroquine can be purchased for about $.10 per course of treatment in many places in Africa, but today it is nearly useless against falciparum malaria, because resistant forms of the parasite are now widespread.

In 2001 the World Health Organization (WHO) declared that artemisinin derivatives should replace chloroquine as the first-line treatment for uncomplicated malaria; these drugs have since proven to be best for severe malaria as well. There is as yet no known resistance of *Plasmodium falciparum* to artemisinin-based drugs. In the preferred ACT formulation, derivatives of the artemisinin compound extracted from the plant are coformulated with another antimalarial drug. The main reason for combining drugs is to inhibit the development of resistance to either drug, much like the approach taken today for HIV/AIDS. Thus, if the parasite begins to develop resistance to artemisinin, the other drug will still kill it, and vice versa. It is much more unlikely that the parasite will develop resistance to both at the same time if the mode of antimicrobial action is different. Widespread use of artemisinin monotherapy—that is, using artemisinin alone—is the biggest threat to the long-term viability of this family of compounds. Artemisinin and another antimalarial can be taken together (in combination) in two ways: as two separate tablets taken at the same time—referred to in this report as a blister pack—or as two drugs in one pill—a coformulation or "fixed-dose combination" referred to here as an ACT. To encourage the use of combination treatment over monotherapy, it is clearly preferable to have both drugs in a single tablet. When two pills must be taken, people often will take just one—in this case, the artemisinin.

Fifty-six countries have adopted ACTs as the first-line or second-line treatment for uncomplicated malaria, and a majority have a general policy in favor of ACTs. And yet only a fraction of the people needing or receiving treatment in Africa will have access to an ACT. ACTs have been effective wherever they have been tried, including highly endemic areas in Africa. However, the supply of ACTs is currently far lower than the need, and the cost is over 20 times that of chloroquine, the most common treatment drug. The side effects of both chloroquine and ACTs are minimal.

The current production situation is evolving. Only one ACT (Coartem) is currently recognized internationally through precertification by WHO, but several others are being manufactured and sold mainly in Asia. Large purchases in Africa—most, if not all, of which employ financing from the Global Fund for AIDS, Tuberculosis and Malaria (Global Fund)—are limited to Coartem or to blister pack combinations, which are also available from a few manufacturers that produce precertified artesunate. Within two years, additional manufacturers are likely to be producing different ACTs.

Nigeria has adopted Coartem as the first-line treatment for uncomplicated malaria, but most people in Nigeria, like those in the rest of Africa, are still using the older, cheaper drugs. Coartem sells for a wholesale price of $2.40 per adult course and about $20 retail in the pharmacy. A blister pack artesunate-amodiaquine combination is sold in Nigeria for $6–$7 per adult course under the brand name Artequin or Artekan.

About 100 million cases of malaria occur each year in Nigeria. However, the market for ACTs is generally calculated on the basis of the "real demand," which takes into account not just need based on malaria incidence, but also national policies and the funding available for purchase by consumers, government, or donor agencies. That figure is about 10 million courses per year, or about a tenth of the estimated medical need. Although the distribution of ACTs has been plagued by shortages since the WHO recommendation on use of ACTs, today the global production capacity is higher than the real (i.e., for which payment is available) demand, which is estimated at between 130 and 220 million treatments, with 30 million currently on order—but still much lower than the medical need. It is clear that ACTs will never be available to the majority of those who need it unless the price is lowered, which can happen realistically only with subsidies of some kind.

Major current sources of funding for government purchases of antimalarials are the Global Fund and the (U.S.) President's Malaria Initiative (PMI). Both of these sources provide funding for all aspects of malaria control, including insecticide-treated bed nets, indoor residual insecticide spraying, as well as drugs. These resource flows are largely (at least to date) limited to the public sector or public sector–like organizations (such as mission hospitals and other health facilities). However, most people do not acquire antimalarials from health facilities; rather, they purchase them through pharmacies or other shops, or in the smallest villages often from drug peddlers. For this reason, the U.S. Institute of Medicine has proposed a "global subsidy" to be applied at the top of the distribution chain (i.e., supranationally) so that it flows through both the public and private sectors, with the aim of an end user price in the range of that charged for chloroquine. The structure of such a plan is currently under development, and the funding, at a level of at least $200 million per year, is still uncertain.

HYPOTHETICAL CASE STUDY

Participants at the April 24-25, 2006, Paris workshop on artemisinin combination therapy created a hypothetical enterprise to produce and market ACTs in Nigeria and the elements of a business plan.

The enterprise. The production and distribution of ACTs encompass several different processes, and it is unlikely, and perhaps not economically feasible, that a single company would carry out all of the tasks and functions involved. However, because the interactions among several partners, if there were such an arrangement, would be a relatively simple part of the establishment of the enterprise, it was convenient to consider the hypothetical enterprise to be a single company that carries out all functions, including production of the artemisinin derivatives through collaboration with an advanced laboratory. The company was given the name Nigerian Anti-malarials Ltd. (henceforth the Company).

Statement of purpose. In view of the morbidity, mortality, and economic burden of malaria control and of the growing resistance of the parasites to the other existing antimalarial drugs, the Company proposes to engage in the profitable production of ACTs that would be sold at an affordable price. Specifically, the Company would engage in

- the cultivation of *A. annua* in Nigeria
- the local extraction and purification of artemisinin
- the production of artemisinin derivatives in collaboration with advanced laboratories
- the local manufacture of ACTs that are globally competitive using current Good Manufacturing Practices (cGMPs).

What is the product? The products would be artemisinin, its derivatives, and artemisinin combination therapies (ACTs).

Who are the customers? For artemisinin, the customers would be local pharmaceutical companies, international organizations, and research laboratories; for artemisinin derivatives, pharmaceutical companies; for ACTs, public consumers, international organizations, NGOs, the government, and pharmaceutical distributors.

Who is the competition? The competition would be provided by several very different sources, including other foreign and local companies making and marketing ACTs in Nigeria and companies using alternative technologies for producing ACTs, offering alternative products, or benefiting from subsidies and grants that might undercut the price charged by the Company.

In a conventional competitive market, other companies, foreign or domestic, might produce a quality product at a lower price, or market it more effectively. But because the process for producing ACTS is highly capital- and technology-intensive, it is unlikely that a new local entrant will emerge that is presently unknown to the Company. Competition from imported foreign products is real, but if the principals of the Company are familiar with the Nigerian market, pharmaceutical community, and environment and are technically well informed, they can expect to remain competitive in the absence of special subsidies to competitors.

China leads in the production of artemisia; it is claimed that 20,000 hectares are under cultivation in Chongqing, much of it by large companies. Altogether, it is expected that more than 150 tons of artemisinin will be produced in 2006. In addition, much of the harvest in China still comes from the wild. Vietnam is a big producer of artemisinin, but yields of artemisia are low compared with what might be expected in Nigeria. India and Brazil are important producers as well. In East Africa, East African Botanicals plans to cultivate 4,000 hectares in Kenya and Tanzania in 2006–2007, but reportedly problems will limit the output. Senegal,

Madagascar, Ghana, and Cameroon have plans to produce ACTs, and they are at about the same stage as Nigeria. A Chinese producer recently informed the Minister of Health of its intention to manufacture ACTs in Nigeria, using artemisinin from local sources.

Among alternative technologies, the most significant in the long term is the development and manufacture of synthetic (small molecule) compounds that act much like artemisinin. Several such compounds are under development by the Medicines for Malaria Venture. Estimates for a marketable synthetic product range from 5 to 10 years, and a transition period of cohabitation is expected. In addition, progress had been made toward the synthesis of artemisinin and related molecules using cell culture. Plant tissue culture to produce artemisia has been tried so far without success.

Some traditional malaria remedies will continue to be available and cheaper than any commercial product. Chloroquine is now the drug of choice for most people. Even artemisinin-based monotherapies may become available, and these will be cheaper than and, in the short term at least, as effective as ACTs. Reducing the availability of monotherapies to discourage parasitic resistance to the drug will be primarily a law enforcement problem and must be left to the government. Meanwhile, counterfeit ACTs have already appeared in Nigeria.

The question of subsidies is more complex and not yet completely resolved. The various donor funds mentioned earlier may be applied in many ways. Other external sources of subsidies, such as foreign companies hoping to enter the Nigerian market, also have offered the government funds for the purchase of malaria drugs. In the most extreme case, funds may be used to import and distribute ACTs at no charge or for a nominal price through public sector institutions, which might seriously distort an incipient ACT market for local producers. (The government has already announced a program of free ACTs for schoolchildren.) Local producers would benefit if the donated ACTs were to be purchased locally at the same time that the price of ACTs was being subsidized in the commercial market. How such an arrangement might actually work is still under discussion. However, a common thread is that all ACTs purchased locally by international sources or exported by national companies would have to be manufactured using WHO-prequalified procedures.

Drugs can be sold on the local market wherever a government permits it. Generally, however, governments rely on WHO certification for current Good Manufacturing Practices, which is the first level for international commerce, to assure quality. To be certified, a company must satisfy criteria related to the purity of materials, processes, inspection of the plant and equipment, quality control, and testing of batches. Several Nigerian companies are cGMP-certified for other drugs. *Prequalification* is a newer and higher level of approval put in place for AIDS, tuberculosis, and malaria drugs. Prequalification requires the use of cGMPs in the process, plus demonstrated product effectiveness against the disease. Prequalification of a product is required for sales supported by the Global Fund and will undoubtedly be a requirement for subsidies on the international market.

That this requirement is more than procedural is testified to by the fact that so far only one company has been prequalified for ACT production: Novartis of Switzerland for Coartem. Even the numerous Chinese manufacturers of other formulations are not prequalified. WHO has announced an effort to increase the number of companies prequalified for ACTs, including generic manufacturers in developing countries.

What are the advantages and challenges for this enterprise in Nigeria? Nigeria has the largest internal market for antimalarials in the world, estimated at 25 percent of the global total. Once Nigerian companies obtain WHO's cGMP certification, they may be able to serve the large market in other countries of West Africa as well. The Nigerian pharmaceutical industry includes several companies that claim the capability to achieve cGMP certification for ACTs. As for the raw materials, recent trials with a variety of *A. annua* cultivars suggest that Nigeria has good growing conditions and may be able to produce multiple harvests annually with abundant yields of artemisinin.

Monotherapies may still be available in Nigeria at prices lower than ACTs, but the Ministry of Health has stated it will approve only ACTs, and will refuse to register any company that produces or imports monotherapies. The licensees that presently have permits for monotherapies will see all those permits expire in two years, and they will not be renewed. Success at producing ACTS would open a niche market for Nigeria in the West African region and create some jobs for the populace.

On the negative side, the inadequacy and unreliability of the infrastructure, including electricity, water, and roads, would be a problem. There are few local suppliers of pharmaceutical chemicals, and many materials would have to be imported, especially the solvents for artemisinin extraction for which no competent local industry currently exists. There is, however, a large market in non-ACT malaria remedies, and a large supplier base for counterfeits and imitations. And, although the government has shown goodwill in addressing some of these problems for the ACTs market, it has a history of not implementing some of its policies.

A single-product company runs the risk that in a few years ACTs could lose their effectiveness or that the compounds could be manufactured more cheaply using genetic engineering. The current research and development pipeline contains some new and novel compounds that may join ACTs as effective antimalarials. However, the next wave approaching the market is based on artemisinin coformulations, which suggests that there will remain a market for artemisinin. An advantageous plan would be to concentrate on malaria for Africa and work on gaining prequalification to benefit from international or national subsidies. Political leaders would favor a local company that is a regional leader in malaria control, carrying out research and training.

Building on the strengths of the Nigerian situation, the Company would have at least two alternate courses of action. It could begin a phased-in process that would lead to a complete supply chain for the manufacture of ACTs within a few years. That process would consist of (1) developing the capability to grow *A. annua*; (2) extracting and purifying artemisinin; (3) manufacturing (or contracting for the manufacture) of the derivative(s); and (4) formulating and packaging ACTs. As a variation, an existing pharmaceutical company could be contracted or taken on as partner to manufacture the product while the Company provides the artemisinin derivative and upgraded equipment. It also could help to secure bank financing.

Instead of trying to compete in the entire production process, the Company could elect to master steps one by one. Because producing a reliable crop of artemisia and extracting the derivative may take a long time, it might be most advantageous to concentrate on the formulation and local marketing of the medicine using imported derivative. Several Nigerian pharmaceutical companies have cGMP-level facilities, and some European companies are offering turnkey factories that are cGMP precertified. At the same time, the best cultivars could be selected and improved, and land could be prepared for cultivation. Seeds could be multiplied while the pharmaceutical plants are under construction. The middle part of the process—extraction, purification, and derivative manufacture—are more difficult and involve imported agents, and the capability must be created or contracted. Meanwhile, a small export market for artemisia leaves, and even for artemisinin, already exists. For a short period, the Company could supply others with artemisinin and receive the derivative in return.

Since the WHO announcement in 2001 recommending ACTs as the first line of treatment, the demand for ACTs has grown exponentially and the price of artemisinin has increased sharply. Even Novartis has had a problem securing materials to support the manufacture of Coartem. As of early 2006, supply and real (paid) demand were about the same, but supply is rising faster than real demand, and the price is expected to fall. The determining factor is the price of leaves, which is presently about $400 a ton at the farm gate. A rule of thumb is that the cost of 1 kilogram of artemisinin is equal to the price of 1 ton of leaves paid to farmers.

Today, the export price of artemisinin is about $400–$600 per kilogram. It appears that any of the component phases of the business can be operated at a profit: growing artemisia, growing plus extracting artemisinin, or manufacturing ACTs.

While in office, President Olusegun Obasanjo created an advisory committee on ACTs. He said that he wanted the ACTs market to be driven by the private sector, and that the government might advance about 40 percent of start-up costs, as was done earlier to encourage the production of cassava by providing cuttings at a subsidized rate. The government might even go so far as to ban the importation of ACTs in the future. The advisory committee's recommendations to President Obasanjo included restricting allocation of permits for growing, extraction, and manufacturing to separate companies, with growers selling to extractors, and these selling to manufacturers. Manufacturers would begin by importing ACTs for marketing in Nigeria, while growers would be planting, multiplying seeds, and providing the harvest to extractors. As of this workshop, the government had not yet responded to these recommendations.

Possible sources of funding. For the process of growing of *A. annua* and producing artemisinin, funding is needed to pay for seeds, to pay farmers for their harvests or field workers to cultivate Company-owned lands, and to pay for drying the leaves and extracting and then purifying the artemisinin. Possible sources of funding are venture capital, banks, individual investors, and government investment in return for equity in the Company.

Several of the workshop participants had had personal experience seeking funding for a Nigerian artemisia project, and they reported a mixed response. Some state governments expressed interest, with the conditions that the enterprise be set up in the state and that a partner be enlisted from the public sector or international donor community. One state was willing to provide the land. Those banks and venture capital companies that were interested insisted on a substantial personal investment by the applicant. One entrepreneur had secured a loan from the U.S. Export-Import Bank (for a different drug), but could not find a local bank that would offer the guaranteed loan at a commercial rate of 25–30 percent.

Some of the large oil companies were interested in helping with such a project in order to overcome unfavorable publicity in Nigeria. But they insisted that the farms and facilities be established in their areas of operation regardless of the suitability for producing artemisinin. The pharmaceutical companies, some of which import artemisinin, appeared to be unresponsive, but meetings continue. The development agencies, such as the U.S. Agency for International Development (USAID) and the Japan

International Cooperation Agency (JICA), were the most forthcoming, offering the promise of grants. USAID has funded a feasibility study, but it also requires a local nonbusiness partner for a grant, possibly a state government.

Producing *A. annua*. Growing *A. annua* could be profitable for farmers, and it could enable them to generate income from a new cash crop with a guaranteed market, at least in the short term. *A. annua* can be grown with other medicinal plants, such as a crop now under study to treat sickle cell disease, and thus would be eligible for the donation of land by the state. Unlike coffee and tea, the major cash crops in the region, *A. annua* is an annual, and its acreage can be adjusted each year in response to the market.

Most experts believe that *A. annua* grows best in the highlands at a cooler temperature. But it is an easy plant to grow and is found nearly everywhere in the wild. With selection and breeding, yields could be high in tropical Nigeria, with the possibility of multiple harvests. In China and Vietnam, the largest producers of *A. annua* leaf at present, farmers have one harvest a year, in summer. But in Brazil, India, Nigeria, and other tropical countries, the combination of high light intensity and no cold season would allow two or more harvests a year for many cultivars. In experience reported from India, a first harvest produced 25 kilograms of artemisinin per hectare, but harvesting leaves after four months while leaving the stems in the field and again harvesting two months later resulted in three or four separate harvests yielding a total of 75 kilograms of artemisinin per hectare. However, harvesting this plant is heavy work, and artemisinin content may vary among harvests. Research should be carried out in each geographic area.

In Nigeria, trials have been carried out only on a small scale in the humid lowlands in the South-South zone (Calabar), and it was necessary to develop a distinct annual cycle to avoid the dry season. Cultivation can begin immediately after the dry season in February and March, then again in July and August. The use of drip irrigation will help to expand the available seasons if necessary for production on a large industrial scale.

The president's advisory committee recommended trials in six states to cover the six climatic zones of the country: Calabar in Cross River state (South-South zone); Enugu in Enugu state (South-East zone); Ota in Ogun state (South-West zone); Plateau (North-Central zone); Bauchi (North-East zone); and Sokoto (North-West zone). The Federal Universities of Technology might also be approached to participate in experimental farms. The University of Calabar already has nine generations of seed, originally from Brazil, China, and elsewhere, created over four years. At present, a cultivar from Brazil gives the highest yield of artemisinin. An accelerated harvesting technique is also used to increase the yield. Leaves are harvested at the peak of their artemisinin content, and the plant is left in the field; the process is then repeated.

The process of extracting artemisinin requires precision in harvest timing and subsequent processing. The plants are not permitted to flower, because, once they do, the artemisinin content drops, and the seeds spread weedy plants that may be difficult to control. This method yields three or four harvests a year. A gram of seeds, which contains 12,000 tiny seeds, or enough to sow half a hectare, can be purchased for $70–$80. The traditional methods of thinning and transplanting were found to be too labor-intensive; better results and higher yields were obtained by broadcasting the seeds. More seeds are required, but the cost of labor is less. The crop is fertilized with the usual nitrogen, phosphorus, potassium (NPK) combinations. Drying can be done in the field in the sun.

The seed garden is maintained separate from the leaf harvest. Artemisinin is bitter to the taste, and cultivar candidates for the seed garden may be selectable by the taste of their leaves. Presently, the price of seeds is rising in Nigeria, and the available supply is rising as well. Because the yield of seeds is high, a 1-hectare seed garden may be enough to serve the Nigerian market, depending on real demand and which ACT is selected. (Different derivatives require differing amounts of artemisinin per course.) It is clear that the supply of seeds will not be a limiting factor until improved hybrid seeds are available.

The wild variety of artemisia growing in Nigeria produces no artemisinin. Nevertheless, it is known and used by people, which suggests it may contain another active compound as yet unknown. Little research has been carried out, because traditionally artemisia has been considered a useless weed. *A. annua* also contains compounds that are of interest to the cosmetics industry, including 3 percent camphor, and so that industry may serve as another market for seeds.

The amount of land required to meet the annual real ACTs demand of Nigeria depends on yield and on the artemisinin derivative used. Table C-1 was developed at the June 2005 Arusha meeting sponsored by the Roll Back Malaria initiative.[1] As shown in the table, about 20,000 hectares will be required to produce 100 million courses of ACTs in Nigeria.

Arrangements with growers. The Company could contract with local farmers to grow *A. annua* where possible to gain the support of the local community and the state. But land tenure systems present a difficulty. Inheritance rules in many regions encourage the breaking up of farms into

[1]Roll Back Malaria initiative, http://www.who.int/malaria/docs/arusha-artemisinin-meeting.pdf.

TABLE C-1 Calculation of Required Planting Area to Meet Nigeria's Need for ACTs

Output	Lower estimate	Average	Upper estimate
Dry leaf yield per hectare	1,000 kg	1,000 kg	1,500 kg
Artemisinin content	0.5%	0.8%	1.0%
Artemisinin yield after processing (40–70%)	2.0 kg (40%)	4.8 kg (60%)	10.5 kg (70%)
Artesunate yield (70%)	1.4 kg	3.4 kg	7.4 kg
Artemether yield (60%)	1.2 kg	2.9 kg	6.3 kg

Note: One ACT adult treatment course contains 0.6 grams of artesunate or 0.48 grams of artemether (for the combination artemether-lumefantrine). Therefore, the number of treatment courses that can be obtained from 1 hectare of artemisia, based on variations in yield, artemisinin content, and recovery, are 2,333–12,250 using artesunate (average 5,600) or 2,500–13,125 using artemether (average 6,000). Producing 100 million adult ACT treatment courses would therefore require 8,160–42,860 hectares for artesunate (average 17, 860 hectares) or 7,620–40,000 hectares for artemether (average 16,670 hectares). With a split of artesunate and artemether, 17,000–18,000 hectares would be required.

Source: "Summary of Working Group B: Artemisinin Extraction, Storage and Quality Assurance," Meeting on the Production of Artemisinin and Artemisinin-Based Combination Therapies, Arusha, Tanzania, June 6–7, 2005, convened by Roll Back Malaria Department, World Health Organization, http://www.who.int/malaria/docs/arusha-artemisinin-meeting.pdf.

small holdings, and small holders must work through cooperatives or be able to benefit from government procurement in order to make economic use of mechanization. The land also may be forested, adding to the cost of land clearing in the humid tropics.

Contract arrangements with growers should provide a guaranteed price that will encourage planting and employment but give no incentive to add filler to the product. In a trial supported by USAID, farmers were given 100 milligrams of seeds, the output of one plant, in the first phase of the trial. They were then required to sell the leaves back to the company in Calabar, and the cost of seeds was subtracted. The price was based on the price of artemisinin; if the international price was $100 per kilogram, farmers were paid $100 per ton of leaves.

As a result of the trial, it is recommended that farmers not be paid for tons of dry biomass. If payment is based on artemisinin content, the analysis must be carried out continuously. The crop should be monitored and collected immediately after harvest and dried elsewhere. If the farmer dries the harvest, he is able to add other material without detection. Collecting the wet biomass and transporting and drying it raise the cost, but the alternative is to supervise the drying.

If land is available to the Company, it would be easier and more economical to contract seasonal labor for planting and harvest, or to contract with an agricultural company and buy back the biomass. The product must be produced following Good Agricultural Practices (GAPs), and it must not be contaminated during the drying process. It can be dried in the sun in for two to three days or in a warehouse; the method of drying does not affect the artemisinin content.

Costs of growing *A. annua*. The cost of cultivation includes:

- land
- land preparation, including equipment purchase and maintenance or lease
- seeds
- labor for planting and harvesting
- extension and training for farmers
- inputs of fertilizer and herbicides, plus irrigation if necessary
- monitoring to ensure quality plants are harvested at the right time.
- drying
- transport of harvested leaves

Extraction and purification of artemisinin. The most commonly used process to extract artemisinin from *A. annua* leaves is solvent extraction. This method is not vulnerable to electrical outage; it can be carried out in the tropics in open buildings with no walls; it operates at relatively low temperatures; it carries no risk of explosion; and it is not protected by patents. Solvents can be recovered and recycled, which lowers the cost and helps protect the environment. The usual solvent is hexane, with an additive to protect against explosion. Other alternatives are petroleum ether, alcohol, or methanol, or the "super-critical solvent process," which may be too difficult to manage for artemisinin. A new method, the BioX process (HFX134A), extracts a higher proportion of the artemisinin, but the materials are expensive and patented, and so royalties must be paid.

Purification can be carried out by crystallization or by chromatography, but both require a heavy investment. Perhaps the best way is to crystallize artemisinin with hexane solvent in a vat and then centrifuge. Impurities can be removed and the crystals vacuum dried to produce pure crystalline artemisinin.

In Nigeria, the solvent must be imported. The Company must check costs and then start with the best proven method available at the time. As the Company becomes more profitable, it might upgrade to new technologies as they become available. Initially, it will cost about 15 percent more to grow, extract, purify, and derive locally than directly importing artemisinin derivatives. To encourage local production, the producers could perhaps be subsidized by the government or by a donor agency.

The cost of equipment depends on whether a steel or glass unit is selected, and whether the fabrication is local. Turnkey plants are available from U.S. and European manufacturers, but the price is high, and the required output based on anticipated demand is still uncertain.

The Calabar group has received a special offer from a manufacturer in China: he offers to import a plant with a capacity of 20 tons per year and operate it for six months for $3 million, apparently below cost. The offer includes another small factory to produce an artemisinin derivative in order to enter the Nigerian market. The demand in Nigeria is about 20 tons, grown on 2,000 hectares, and one extraction plant in one location should be sufficient to start. Initially, the artemisinin product could be sold to Novartis or another established firm, and eventually a partnership could be set up that returns the derivatives for formulation.

As of early 2006, the price of artemisinin was about $600 per kilogram, but within one or two years it is likely to drop to about $350. The cost of production depends on the price of leaf and yield of artemisinin, and so the local costs of production would have to be monitored against the cost of importation. The price of the derivative artesunate is $970 per kilogram.

If the Chinese offer were accepted, about six months would be required to import, transport, install, and test all equipment. The equipment is not modular, and some elements can be used for other processes. Similarly, six months would be required to grow and dry the raw material. However, to anticipate contingencies nine months should be allowed, and so the process must be started in May to meet the growing season.

Regulatory requirements. Minor environmental issues must be taken into account for both the cultivation stage and the production process. *A. annua* is a weed, and care must be taken in dispersing seeds so that plants do not take over neighboring fields. An environmental impact assessment must be carried out, and a permit from the Ministry of the Environment and Housing will be required.

In the cultivation and production of artemisinin, no toxic effects on humans or animals have been identified, but organic reactants should always be used with care. Solvents are recycled, but the effluents generated may present environmental problems. A mini-effluent plant should be built into the extraction plant. Leaf waste will also be produced, and the dry residue will contain solvent. This residue will be difficult to treat and may have to be burned, and the heat and fumes must be dispersed.

The purification process also produces a waxy substance with no known application that must be discarded.

The Company will have to be registered with the drug regulation authority and the corporate affairs commission of the state ministry of commerce and industry.

Costs of extraction and purification. The artemisinin production costs will include the following:

- A turnkey plant. The Chinese offering can be purchased for $3 million.
- A technical-grade solvent. Ten liters of solvent are required for 1 kilogram of plant material. One liter of hexane costs $.50, and so the solvent required for 1 kilogram of leaves would cost about $5. The solvent is normally recycled, with about 15 percent losses in recycling. The cost of the solvent per kilogram of leaves would therefore be about $.75.
- Normal operating expenses
- Security
- Devices to prevent damage from lightning in a rainy climate like that of Calabar
- Power
- Labor
- Insurance.

Manufacture of ACTs. It should be possible for a company to operate under cGMP conditions certified by WHO, thereby allowing the product to be sold in Nigeria and other countries willing to accept the import. But without prequalification, the product would not be eligible for international purchase (e.g., using funds from WHO or the Global Fund), nor for a subsidy if that option materialized. Such a situation would create serious competition from subsidized medicines, either imported or from other local companies.

Choice of ACT. All three of the common agents—dihydroartemisinin (DHA), artesunate, and artemether—could be produced in Nigeria, but the clear choice is artesunate or DHA, because artemether plus lumefantrine (AL), the formulation presently adopted by the Nigerian government, is under patent to Novartis as Coartem until 2011. In Nigeria, under the agreement with WHO the price of Coartem is $2.40 per course of treatment. Worldwide, it is available for about 10 times that price in the public sector and for up to $40 in the private sector, where many people buy their antimalarials. ACTs that are less expensive to manufacture than Coartem (in part because the companion drugs are less expensive) are under development and may be available in a few years for a wholesale price of $1 to the public sector and $2–$3 in the private sector. The Drugs for Neglected Diseases initiative (DNDi) is evaluating several different combinations and sponsoring clinical trials in other countries in Africa, and it will prepare an assessment of all trials.

Technologies used. The $3 million price of the Chinese offer in Calabar includes a small additional factory that would make the derivative alongside the extraction of artemisinin. The process is not complex and can be done in Nigeria.

In one example, artemisinin can be converted to artemether using a well-known two-step process in which, first, ketone is reduced to alcohol at normal temperatures (which gives a 97 percent yield in the laboratory), and then the alcohol is converted to beta ether for artemether. At this point, it may be expedient to send the material to an advanced laboratory to make the derivative at additional cost.

Most of the equipment required for producing the product can be purchased off the shelf. This equipment includes a mill, dryer, granulator, packaging tools, and the analytical laboratory. Coformulation will require special equipment, and the process will be spelled out in the license and dossiers provided by WHO for prequalification.

Site. ACTs production should be located in an urban center with adequate infrastructure, including electric power, roads, and a water supply. It could be placed near the cultivators, in a city such as Calabar, but that location would mean incurring extra transportation cost to Lagos, the major commercial center in Nigeria where most of the pharmaceutical companies are located. Calabar does, however, offer the possibility of support from the state government. Because malaria is a national concern, treatment therapies are not subject to federal duties, and the equipment and raw materials needed for the production of ACTs should attract no duties in any port of entry into Nigeria.

Licensing. To put a drug on the market in Nigeria, a company must be registered with the National Agency for Food, Drug Administration, and Control (NAFDAC). The Company also must register with the state ministry of commerce and industry and receive approval from the Ministry of the Environment and Housing. Finally, it must register with the Pharmaceutical Society of Nigeria, a nongovernmental organization (NGO).

The product itself also must be registered with NAFDAC, and documentation must be provided on clinical trials. For an existing formulation, the documentation must demonstrate manufacturing capability and equivalency with the drug approved elsewhere. In principle, new clinical trials might be required, but the Nigerian government will not require new trials for ACTs approved elsewhere.

Marketing. The trade name for the product will be an important factor in the marketing plan. Nigerians are accustomed to taking chloroquine and stopping the treatment when and if improvement is felt. The marketing campaign should promote the idea of a cure within a fixed number of days if the course is completed. There would, of course, be different dosages for adults and children. New formulations for adults are likely to require two tablets once a day for three days. Coartem is a course of 24 pills over three days. Therefore, the trade name might incorporate the idea of a three-day cure, and the campaign might emphasize the drug's advantage over chloroquine in that the treatment need be taken only in one course for a cure.

The product should be marketed to NGOs, wholesalers, and member countries of the Economic Community of West African States (ECOWAS), and through educational programs in schools, with emphasis on completing therapy. Presently, much malaria advertising is placed on billboards, but the government has indicated its intention to prohibit billboard advertising for antimalarial drugs. The health system should be utilized as well as possible. Trained field workers could work to ensure compliance.

Another marketing strategy could involve age-specific packaging. Different dosages are appropriate for different age groups, and the packages could contain diagrams showing frequency and time of administration. Children under five may be treated presumptively without diagnosis.

Implementation plan. There is already considerable momentum toward growing *A. annua* and manufacturing ACTs in Nigeria. Officers of three companies engaged in these activities were members of the workshop panel, and other Nigerian and foreign companies have demonstrated interest.

Financing. It is quite likely that an established pharmaceutical company would play a large role in the manufacture of ACTs, whether as a partner of a new company or as the prime entity that will subcontract the growing of artemisia and the extraction of artemisinin. The established company would likely have to dedicate some of its own capital to the venture in order to qualify for financing from a commercial bank or an international financing agency such as the African Development Bank or the International Finance Corporation. The probable cost of the initiative would be in the range $10–$15 million. For a new company, venture capital companies, particularly the newer "social" venture capital companies, might be interested.

Cost of a cGMP facility to manufacture ACTs (contributed by Prof. Krishna Kumar, Howard University, Washington, DC). The costs mentioned here would apply to the United States, and so they may differ from the costs in Nigeria, especially where transport and import fees must be added.

There are three options for ACT formulation:

1. *Direct compression.* Involves mixing the raw material and compressing it into tablets. This process would require (1) a V blender with a capacity of 5 cubic feet for a 30–50 kilogram batch, at an approximate cost of $25,000 for used and $50,000 for new; and (2) a tablet press for a 30–50 kilogram batch, at a cost of $30,000 for a used mini-press and $60,000 for a new one, or a cost of $45,000 for a used beta press and $120,000 for a new one.

2. *Granulation followed by compression.* To the costs in option 1 must be added the cost of granulation and drying equipment. The cost of a granulator and tray drier is $50,000 used and $100,000 new.

3. *Powder mixing followed by capsulation.* This process would require a V blender with a capacity of 5 cubic feet for a 30–50 kilogram batch, at an approximate cost $25,000 for used and $50,000 for new; and a capsule-filling machine at a cost of $30,000 used and $50,000 new for a table-top model with a capacity of 20,000 capsules per hour, or $65,000 used and $110,000 new for a floor model with a capacity of 40,000 capsules per hour. This option may also require granulation, depending on the formulation and excipients used. In that case, the cost of granulation equipment, as in option 2, would have to be included.

Some ACTs would also require film coating to fill aesthetic and stability requirements. The cost of film-coating equipment for a 24-inch pan for 30 kilograms is $30,000 used and $90,000 new, and for a 48-inch pan for 75–100 kilograms is $75,000 used and $150,000 new.

In addition to these equipment costs, a minimum packaging suite based on bottles and cartons would have to be included. A basic model consisting of a counter, conveyer, filler and capper, labeler, printer, and ability to pack cartons could be assembled for about $125,000. A new machine could cost up to $250,000 depending on the options. Some of the packaging costs may be reduced by making the operation more labor-intensive, but maintaining GMP conditions and quality could be difficult.

To follow cGMP procedures, it also will be necessary to include some in-process quality control equipment such as instruments to measure hardness ($3,500 used, $7,000 new), friability ($3,000 used, $7,000–$10,000 new), tap density ($3,000 used, $5,000 new), and disintegration ($5,500 new). Accessories such as a metal detector ($10,000), sonic sifter ($5,000), and dust collector ($3,000–$10,000) also would be needed.

If a building that meets cGMP requirements for temperature and humidity control, air handling, and water is not available, the cost of designing and constructing such a building must be added.

Government role. The Nigerian government could promote the manufacture and distribution of ACTs in several ways. It could provide seed money for new start-ups or grants to encourage pharmaceutical companies to enter the market and to help companies to achieve precertification. A donation of land for growing artemisia would shorten the timetable considerably.

It is important that the government provide political goodwill and an enabling environment. Banning counterfeits, clones, and low-quality products from the market would both help the new manufacturer and protect the public at the same time. Public campaigns calling for using insecticide-treated bed nets, cleaning up standing water sources, and taking the correct medicine, and that promote ACTs and oppose monotherapy would be helpful both to the public and to the suppliers of ACTs. Finally, a program of public education in the schools on the avoidance and control of malaria should be cost-effective and successful, just as similar programs on HIV/AIDS and smoking have been in other places.

The government could take the following specific actions to encourage ACTs production:

- waive any existing duties on imported equipment, raw materials, solvents, and other materials involved in ACTs production
- donate land for farming of *A. annua*
- enforce a ban on antimalarial monotherapies for uncomplicated malaria
- provide an administrative infrastructure
- assist in training and research
- purchase media advertising to promote effective malaria therapy and completion of courses of treatment
- pass legislation to require the distribution and use of ACTs in all government schools, hospitals, and agencies.

Role of the science academies. The U.S. National Academies and the Nigerian Academy of Science could collaborate by designing and preparing the terms of reference for an educational program on malaria control for the schools. The Nigerian academy is prepared to play a strong advocacy role within Nigeria for the local production of ACTs and the exclusive use of ACTs for malaria control. It might also help interested companies become informed about the requirements for cGMP certification and WHO prequalification and stay abreast of developments in malaria drug financing.

PARTICIPANTS, ARTEMISININ-BASED MALARIA THERAPY WORKSHOP, PARIS

Edith Ajaiyeoba	University of Ibadan
Nancy Bearg	Consultant
Ebiamadon Andi Brisibe	University of Calabar, Molecular Bio/Sciences Ltd.
Yves Champey	Drugs for Neglected Diseases initiative
Catherine Falade	University of Ibadan
Hellen Gelband	Institute of Medicine, U.S. National Academies
Michael Greene	U.S. National Academies
Jean-Marie Kindermans	Médicins sans Frontières
G. B. Ogunmola	Nigerian Academy of Science, University of Ibadan
Folasade Olaiya-Segun	Neimeth International Pharmaceuticals
Ramesh Pandey	XeChem Inc.
Jacques Pilloy	Office Technique d'Études et de Coopération Internationales
George Tyler	Drugs for Neglected Diseases initiative